262
Topics in Current Chemistry

Editorial Board:
V. Balzani · A. de Meijere · K. N. Houk · H. Kessler · J.-M. Lehn
S. V. Ley · S. L. Schreiber · J. Thiem · B. M. Trost · F. Vögtle
H. Yamamoto

Topics in Current Chemistry

Recently Published and Forthcoming Volumes

Molecular Machines
Volume Editor: Kelly, T. R.
Vol. 262, 2006

Immobilisation of DNA on Chips II
Volume Editor: Wittmann, C.
Vol. 261, 2005

Immobilisation of DNA on Chips I
Volume Editor: Wittmann, C.
Vol. 260, 2005

Prebiotic Chemistry
From Simple Amphiphiles to Protocell Models
Volume Editor: Walde, P.
Vol. 259, 2005

Supramolecular Dye Chemistry
Volume Editor: Würthner, F.
Vol. 258, 2005

Molecular Wires
From Design to Properties
Volume Editor: De Cola, L.
Vol. 257, 2005

Low Molecular Mass Gelators
Design, Self-Assembly, Function
Volume Editor: Fages, F.
Vol. 256, 2005

Anion Sensing
Volume Editor: Stibor, I.
Vol. 255, 2005

Organic Solid State Reactions
Volume Editor: Toda, F.
Vol. 254, 2005

DNA Binders and Related Subjects
Volume Editors: Waring, M. J., Chaires, J. B.
Vol. 253, 2005

Contrast Agents III
Volume Editor: Krause, W.
Vol. 252, 2005

Chalcogenocarboxylic Acid Derivatives
Volume Editor: Kato, S.
Vol. 251, 2005

New Aspects in Phosphorus Chemistry V
Volume Editor: Majoral, J.-P.
Vol. 250, 2005

Templates in Chemistry II
Volume Editors: Schalley, C. A., Vögtle, F., Dötz, K. H.
Vol. 249, 2005

Templates in Chemistry I
Volume Editors: Schalley, C. A., Vögtle, F., Dötz, K. H.
Vol. 248, 2004

Collagen
Volume Editors: Brinckmann, J., Notbohm, H., Müller, P. K.
Vol. 247, 2005

New Techniques in Solid-State NMR
Volume Editor: Klinowski, J.
Vol. 246, 2005

Functional Molecular Nanostructures
Volume Editor: Schlüter, A. D.
Vol. 245, 2005

Natural Product Synthesis II
Volume Editor: Mulzer, J.
Vol. 244, 2005

Natural Product Synthesis I
Volume Editor: Mulzer, J.
Vol. 243, 2005

Molecular Machines

Volume Editor: T. Ross Kelly

With contributions by
V. Balzani · J.-P. Collin · A. Credi · J. M. Fernández · B. Ferrer
A. H. Flood · M. A. Garcia-Garibay · V. Heitz · S. D. Karlen
E. R. Kay · D. A. Leigh · T. F. Magnera · J. Michl · N. N. P. Moonen
J.-P. Sauvage · S. Silvi · J. F. Stoddart · M. Venturi

The series *Topics in Current Chemistry* presents critical reviews of the present and future trends in modern chemical research. The scope of coverage includes all areas of chemical science including the interfaces with related disciplines such as biology, medicine and materials science. The goal of each thematic volume is to give the nonspecialist reader, whether at the university or in industry, a comprehensive overview of an area where new insights are emerging that are of interest to a larger scientific audience.

As a rule, contributions are specially commissioned. The editors and publishers will, however, always be pleased to receive suggestions and supplementary information. Papers are accepted for *Topics in Current Chemistry* in English.

In references *Topics in Current Chemistry* is abbreviated Top Curr Chem and is cited as a journal.

Visit the TCC content at springerlink.com

ISSN 0340-1022
ISBN-10 3-540-28501-6 Springer Berlin Heidelberg New York
ISBN-13 978-3-540-28501-4 Springer Berlin Heidelberg New York
DOI 10.1007/b105501

This work is subject to copyright. All rights are reserved, whether the whole or part of the material is concerned, specifically the rights of translation, reprinting, reuse of illustrations, recitation, broadcasting, reproduction on microfilm or in any other way, and storage in data banks. Duplication of this publication or parts thereof is permitted only under the provisions of the German Copyright Law of September 9, 1965, in its current version, and permission for use must always be obtained from Springer. Violations are liable for prosecution under the German Copyright Law.

Springer is a part of Springer Science+Business Media

springer.com

© Springer-Verlag Berlin Heidelberg 2005
Printed in Germany

The use of registered names, trademarks, etc. in this publication does not imply, even in the absence of a specific statement, that such names are exempt from the relevant protective laws and regulations and therefore free for general use.

Cover design: *Design & Production* GmbH, Heidelberg
Typesetting and Production: LE-TEX Jelonek, Schmidt & Vöckler GbR, Leipzig

Printed on acid-free paper   02/3141 YL – 5 4 3 2 1 0

Volume Editor

Professor T. Ross Kelly
Department of Chemistry
Boston College
Merke Chemistry Center
2609 Beacon St.
Chestnut Hill, MA 02467, USA
ross.kelly@bc.edu

Editorial Board

Prof. Vincenzo Balzani
Dipartimento di Chimica „G. Ciamician"
University of Bologna
via Selmi 2
40126 Bologna, Italy
vincenzo.balzani@unibo.it

Prof. Dr. Armin de Meijere
Institut für Organische Chemie
der Georg-August-Universität
Tammanstr. 2
37077 Göttingen, Germany
ameijer1@uni-goettingen.de

Prof. Dr. Kendall N. Houk
University of California
Department of Chemistry and
Biochemistry
405 Hilgard Avenue
Los Angeles, CA 90024-1589
USA
houk@chem.ucla.edu

Prof. Dr. Horst Kessler
Institut für Organische Chemie
TU München
Lichtenbergstraße 4
86747 Garching, Germany
kessler@ch.tum.de

Prof. Jean-Marie Lehn
ISIS
8, allée Gaspard Monge
BP 70028
67083 Strasbourg Cedex, France
lehn@isis.u-strasbg.fr

Prof. Steven V. Ley
University Chemical Laboratory
Lensfield Road
Cambridge CB2 1EW
Great Britain
Svl1000@cus.cam.ac.uk

Prof. Stuart L. Schreiber
Chemical Laboratories
Harvard University
12 Oxford Street
Cambridge, MA 02138-2902
USA
sls@slsiris.harvard.edu

Prof. Dr. Joachim Thiem
Institut für Organische Chemie
Universität Hamburg
Martin-Luther-King-Platz 6
20146 Hamburg, Germany
thiem@chemie.uni-hamburg.de

Prof. Barry M. Trost

Department of Chemistry
Stanford University
Stanford, CA 94305-5080
USA
bmtrost@leland.stanford.edu

Prof. Dr. F. Vögtle

Kekulé-Institut für Organische Chemie
und Biochemie
der Universität Bonn
Gerhard-Domagk-Str. 1
53121 Bonn, Germany
voegtle@uni-bonn.de

Prof. Dr. Hisashi Yamamoto

Department of Chemistry
The University of Chicago
5735 South Ellis Avenue
Chicago, IL 60637
USA
yamamoto@uchicago.edu

Topics in Current Chemistry
Also Available Electronically

For all customers who have a standing order to Topics in Current Chemistry, we offer the electronic version via SpringerLink free of charge. Please contact your librarian who can receive a password or free access to the full articles by registering at:

springerlink.com

If you do not have a subscription, you can still view the tables of contents of the volumes and the abstract of each article by going to the SpringerLink Homepage, clicking on "Browse by Online Libraries", then "Chemical Sciences", and finally choose Topics in Current Chemistry.

You will find information about the

– Editorial Board
– Aims and Scope
– Instructions for Authors
– Sample Contribution

at springeronline.com using the search function.

Preface

The quest for ever smaller devices has stimulated a transforming reassessment of how such objects might be achieved. Initially, the strategy was to shrink and shrink again macroscopic constructs to arrive at microscopic analogs. Perhaps the ultimate embodiment of this so-called "top down" approach has been manifested in integrated circuits. For thirty years Moore's Law in its various forms (basically, that the amount of space required on a chip for a unit of computing power–or for a transistor–decreases by a factor of two every two years; put another way, a 2005 chip can carry out $2^{15} = 33\,000$ times as many operations as a same-sized chip manufactured in 1975). It has long been argued that physical limits, arising primarily from the laws of optics, will eventually put an end to further progress using the top-down approach.

Such limits do not afflict the "bottom up" approach, where one starts with individual atoms and builds from there. If the ultimate in miniaturization is sought, then the goal is molecules, assemblages of atoms. The design, construction and study of molecules are the special purview of chemistry. Chemists have plied their trade for many decades, but it is only recently that they have directed their attention to achieving molecular scale machines. Consequently, compared to the top down strategy, the bottom up approach is still in its infancy. The opportunities as well as the challenges are enormous. The chapters in this volume all describe bottom up strategies and chronicle cutting-edge advances from several of the world's leading laboratories engaged in the development of molecular machines. One cannot help but be dazzled by the manifest ingenuity on display: It gives a whole new meaning to the term chemical engineering.

It has been a delight as well as a privilege to work with an outstanding group of contributing authors and I thank them for all their efforts. I also acknowledge the National Institutes of Health for support of my laboratory's work in this area (grant number GM56262).

Chestnut Hill, September 2005   T. Ross Kelly

Contents

Artificial Molecular Motors and Machines:
Design Principles and Prototype Systems
V. Balzani · A. Credi · B. Ferrer · S. Silvi · M. Venturi 1

Transition-Metal-Complexed Catenanes and Rotaxanes in Motion:
Towards Molecular Machines
J.-P. Collin · V. Heitz · J.-P. Sauvage 29

Altitudinal Surface-Mounted Molecular Rotors
T. F. Magnera · J. Michl 63

Towards a Rational Design of Molecular Switches and Sensors
from their Basic Building Blocks
N. N. P. Moonen · A. H. Flood · J. M. Fernández · J. F. Stoddart 99

Hydrogen Bond-Assembled Synthetic Molecular Motors and Machines
E. R. Kay · D. A. Leigh 133

Amphidynamic Crystals:
Structural Blueprints for Molecular Machines
S. D. Karlen · M. A. Garcia-Garibay 179

Author Index Volumes 251–262 229

Subject Index 235

Contents of *Structure and Bonding, Vol. 99*

Molecular Machines and Motors

Volume Editor: J.-P. Sauvage
ISBN: 3-540-41382-0

Single Molecular Rotor at the Nanoscale
C. Joachim, J. K. Gimzewski

Rotary Motion in Single-Molecule Machines
T. R. Kelly, J. P. Sestelo

**Molecular Machines and Motors Based
on Transition Metal-Containing Catenanes and Rotaxanes**
L. Raehm, J.-P. Sauvage

**Molecular Movements and Translocations Controlled
by Transition Metals and Signaled by Light Emission**
V. Amendola, L. Fabbrizzi, C. Mangano, P. Pallavicini

**Molecular Hysteresis by Linkage Isomerizations Induced
by Electrochemical Processes**
M. Sano

**Switchable Molecular Devices:
From Rotaxanes to Nanoparticles**
J. Liu, M. Gómez-Kaifer, A. E. Kaifer

**Molecular-Level Artificial Machines Based
on Photoinduced Electron-Transfer Processes**
R. Ballardini, V. Balzani, A. Credi, M. T. Gandolfi, M. Venturi

Computing at the Molecular Level
A. R. Pease, J. F. Stoddart

**Molecular Memory and Processing Devices
in Solution and on Surfaces**
A. N. Shipway, E. Katz, I. Willner

Artificial Molecular Motors and Machines: Design Principles and Prototype Systems

Vincenzo Balzani · Alberto Credi (✉) · Belen Ferrer · Serena Silvi · Margherita Venturi

Dipartimento di Chimica "G. Ciamician", Università di Bologna, via Selmi 2, 40126 Bologna, Italy
alberto.credi@unibo.it

1	Basic Principles	2
1.1	Introduction	2
1.2	The Bottom-Up (Supramolecular) Approach to Nanodevices	3
1.3	Characteristics of Molecular Motors and Machines	5
1.3.1	Energy Supply	5
1.3.2	Other Features	6
1.4	Natural Molecular Motors and Machines	7
1.5	Rotaxanes and Catenanes as Artificial Molecular Machines	8
2	Prototypes	11
2.1	Linear Motions in Rotaxanes	11
2.2	An Acid-Base Controlled Molecular Shuttle	12
2.3	A Molecular Elevator	13
2.4	An Autonomous Light-Powered Molecular Motor	16
2.5	Systems Based on Catenanes	19
3	Conclusion and Perspectives	22
	References	24

Abstract A molecular machine can be defined as an assembly of a discrete number of molecular components (that is, a supramolecular structure) designed to perform a function through the mechanical movements of its components, which occur under appropriate external stimulation. Hence, molecular machines contain a motor part, that is a device capable of converting energy into mechanical work. Molecular motors and machines operate via nuclear rearrangements and, like their macroscopic counterparts, are characterized by the kind of energy input supplied to make them work, the manner in which their operation can be monitored, the possibility to repeat the operation at will, i.e., establishing a cyclic process, the time scale needed to complete a cycle of operation, and the performed function. Owing to the progress made in several branches of Chemistry, and to the better understanding of the operation mechanisms of molecular machines of the biological world, it has become possible to design and construct simple prototypes of artificial molecular motors and machines. The extension of the concept of machine to the molecular level is of great interest not only for basic research, but also for the growth of nanoscience and the development of nanotechnology. We will illustrate some basic features and design principles of molecular machines, and we will describe a few recent examples of artificial systems, based on rotaxanes, catenanes and related species, taken from our own research.

Keywords Catenane · Electron transfer · Photochemistry · Rotaxane · Supramolecular chemistry

Abbreviations
AMH   ammonium center
BPM   4,4'-bipyridinium unit
CT   charge transfer
DB24C8   dibenzo[24]crown-8
DON   1,5-dioxynaphthalene unit
SCE   saturated calomel electrode
TTF   tetrathiafulvalene unit

1
Basic Principles

1.1
Introduction

Movement is one of life's central attributes. Nature provides living systems with complex molecules called motor proteins which work inside a cell like ordinary machines built for everyday needs. The development of civilization has always been strictly related to the design and construction of devices, from wheel to jet engine, capable of facilitating movement and travel. Presently, the miniaturization race leads scientists to investigate the possibility of designing and constructing motors and machines at the nanometer scale, i.e. at the molecular level. Chemists, by the nature of their discipline, are able to manipulate atoms and molecules and are therefore in an ideal position to develop bottom-up strategies for the construction of nanoscale devices.

Natural molecular motors are extremely complex systems; their structures and detailed working mechanisms have been elucidated only in a few cases and any attempt to construct systems of such a complexity by using a molecular approach would be hopeless. What can be done, at present, in the field of artificial molecular motors is (i) to learn the design principles by constructing simple prototypes consisting of a few molecular components, capable of moving in a controllable way; and (ii) to investigate the challenging problems posed by interfacing artificial molecular devices with the macroscopic world, particularly as far as energy supply and information exchange are concerned. Surely, the study of motion at the molecular level is a fascinating topic from the viewpoint of basic research and a promising field for novel applications.

In the first section of this chapter we shall introduce the concepts of molecular motors and machines, and we will describe the bottom-up (i.e.

supramolecular) approach to their construction. We will then discuss the characteristics of molecular motors, with reference to those of macroscopic ones. The second section deals with artificial molecular motors and machines based on rotaxanes, catenanes and related species. For space reasons, we shall only describe a few recent examples taken from our own research. Finally, we will critically discuss some perspectives and limitations of these kinds of systems.

1.2
The Bottom-Up (Supramolecular) Approach to Nanodevices

In everyday life we make extensive use of devices. A *device* is something invented and constructed for a special purpose. More specifically, it is an assembly of components designed to achieve a specific *function*, resulting from the cooperation of the (simple) *acts* performed by each component (Fig. 1a). A *machine* is a particular type of device in which the component parts display changes in their relative positions as a result of some external stimulus.

Depending on the purpose of its use, a device can be very big or very small. In the last fifty years, progressive miniaturization of the components employed for the construction of devices and machines has resulted in outstanding technological achievements, particularly in the field of information processing. A common prediction is that further progress in miniaturization

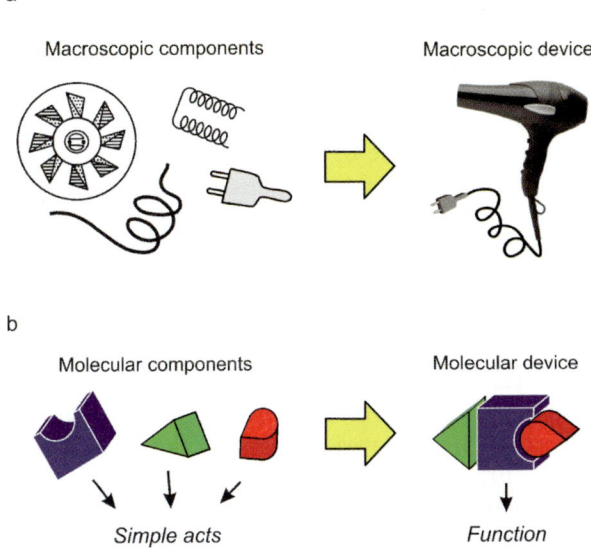

Fig. 1 Extension of the concept of macroscopic device to the molecular level

will not only decrease the size and increase the power of computers, but could also open the way to new technologies in the fields of medicine, environment, energy, and materials.

Until now miniaturization has been pursued by a large-downward (top-down) approach, which is reaching practical and fundamental limits (presumably ca. 50 nanometers) [1]. Miniaturization, however, can be pushed further since "there is plenty of room at the bottom", as Richard Feynman stated in a famous talk to the American Physical Society in 1959 [2, 3]. The key sentence of Feynman's talk was the following: *"The principle of physics do not speak against the possibility of manoeuvring things atom by atom"*. The idea of the "atom-by-atom" bottom-up approach to the construction of nanoscale devices and machines, however, which was so appealing to some physicists [4, 5] did not convince chemists who are well aware of the high reactivity of most atomic species and of the subtle aspects of the chemical bond. Chemists know [6] that atoms are not simple spheres that can be moved from one place to another at will. Atoms do not stay isolated; they bind strongly to their neighbors and it is difficult to imagine that the atoms can be taken from a starting material and transferred to another material.

In the late 1970s a new branch of chemistry, called *supramolecular chemistry*, emerged and expanded very rapidly. In the frame of research on supramolecular chemistry, the idea began to arise in a few laboratories [7–9] that molecules are much more convenient building blocks than atoms to construct nanoscale devices and machines (Fig. 1b). The main reasons for this idea are: (i) molecules are stable species, whereas atoms are difficult to handle; (ii) nature starts from molecules, not from atoms, to construct the great number and variety of nanodevices and nanomachines that sustain life; (iii) most of the laboratory chemical processes deal with molecules, not with atoms; (iv) molecules are objects that exhibit distinct shapes and carry device-related properties (e.g. properties that can be manipulated by photochemical and electrochemical inputs); (v) molecules can self-assemble or can be connected to make larger structures. In the same period, research on molecular electronic devices began to flourish [10–12].

In the following years supramolecular chemistry grew very rapidly [13–16] and it became clear that the bottom-up approach based on molecules opens virtually unlimited possibilities concerning the design and construction of artificial molecular devices and machines. Recently, the concept of molecules as nanoscale objects exhibiting their own shape, size, and properties has been confirmed by new, very powerful techniques, such as single-molecule fluorescence spectroscopy and the various types of probe microscopies, capable of visualizing [17–19] or manipulating [20–22] single molecules, and even investigating bimolecular chemical reactions at the single molecule level [23].

Much of the inspiration to construct molecular devices and machines comes from the outstanding progress of molecular biology that has begun to reveal the secrets of the natural nanodevices which constitute the material base of life [24]. The bottom-up construction of devices as complex as those present in Nature is, of course, an impossible task. Therefore, chemists have tried to construct much simpler systems, without mimicking the complexity of the biological structures. In the last few years the synthetic talent, that has always been the most distinctive feature of chemists, combined with a device-driven ingenuity evolved from chemists' attention to functions and reactivity, have led to outstanding achievements in this field [25–29].

1.3
Characteristics of Molecular Motors and Machines

A *molecular motor* can be defined as an assembly of a discrete number of molecular components designed to perform mechanical-like movements under the control of appropriate energy inputs. This definition excludes the molecular motions caused simply by thermal energy [30–36]. The words *motor* and *machine* are often used interchangeably when referred to molecular systems. It should be recalled, however, that a motor converts energy into mechanical work, while a machine is a device, usually containing a motor component, designed to accomplish a function. Molecular motors and machines operate via electronic and/or nuclear rearrangements and make use of thermal fluctuations (Brownian motion) [37–39]. Like the macroscopic counterparts, they are characterized by (i) the kind of energy input supplied to make them work; (ii) the type of motion (linear, rotatory, oscillatory, …) performed by their components; (iii) the way in which their operation can be monitored; (iv) the possibility to repeat the operation at will (cyclic process); and (v) the time scale needed to complete a cycle. According to the view described above, an additional and very important distinctive feature of a molecular machine with respect to a molecular motor is (vi) the function performed [26].

1.3.1
Energy Supply

The problem of the energy supply to make artificial molecular motors work [point (i)] is of the greatest importance [40]. The most obvious way to supply energy to a chemical system is through an exergonic chemical reaction. In the previously mentioned address [2, 3] to the American Physical Society, Richard Feynman observed: *"An internal combustion engine of molecular size is impossible. Other chemical reactions, liberating energy when cold, can be used instead"*. This is exactly what happens in our body, where the chemical energy supplied by food is used in a long series of slightly exergonic reactions to power the biological machines that sustain life.

If an artificial molecular motor has to work by inputs of chemical energy, it will need the addition of fresh reactants ("fuel") at any step of its working cycle, with the concomitant formation of waste products. Accumulation of waste products, however, will compromise the operation of the device unless they are removed from the system, as happens in our body as well as in macroscopic internal combustion engines. The need to remove waste products introduces noticeable limitations in the design and construction of artificial molecular motors based on chemical fuel inputs.

Chemists have long since known that photochemical and electrochemical energy inputs can cause the occurrence of *endergonic* and *reversible* reactions. In recent years, the outstanding progress made by supramolecular photochemistry [41–43] and electrochemistry [44, 45] has thus led to the design and construction of molecular machines powered by light or electrical energy, which work without the formation of waste products. In the case of photoexcitation, the commonly used endergonic and reversible reactions are isomerization and redox processes. In the case of electrochemical energy inputs, the induced endergonic and reversible reactions are, of course, heterogeneous electron transfer processes. Photochemical and electrochemical techniques offer further advantages, since lasers provide the opportunity of working in very small spaces and very short time domains, and electrodes represent one of the best ways to interface molecular-level systems with the macroscopic world.

A very important feature of molecular motors, related to energy supply [point (i)] and cyclic operation [point (iv)], is their capability to exhibit an *autonomous* behavior; that is, to keep operating, in a constant environment, as long as the energy source is available. Natural motors are autonomous, but the vast majority of the artificial molecular motors reported so far are *not autonomous* since, after the mechanical movement induced by a given input, they need another, opposite input to reset.

Needless to say, the operation of a molecular machine is accompanied by partial conversion of free energy into heat, regardless of the chemical, photochemical, and electrochemical nature of the energy input.

1.3.2
Other Features

The motions performed by the component parts of a molecular motor [point (ii)] may imply rotations around covalent bonds or the making and breaking of intercomponent non-covalent bonds, as we shall see later on.

In order to control and monitor the device operation [point (iii)], the electronic and/or nuclear rearrangements of the component parts should cause readable changes in some chemical or physical property of the system. In this regard, photochemical and electrochemical techniques are very useful since both photons and electrons can play the dual role of *writing* (i.e. causing a change in the system) and *reading* (i.e. reporting the state of the system). Luminescence spectroscopy, in particular, is a most valuable reading technique since it is easily accessible and offers good sensitivity and selectivity, along with the possibility of time-resolved studies.

The operation time scale of molecular machines [point (v)] can range from microseconds to seconds, depending on the type of rearrangement and the chemical nature of the components involved.

Finally, as far as point (vi) is concerned, the functions that can be performed by exploiting the movements of the component parts in molecular motors and machines are various and, to a large extent, still unpredictable. In natural systems the molecular motions are always aimed at obtaining specific functions, e.g., catalysis, transport, gating. As will be described in the following sections, the mechanical movements taking place in molecular machines, and the related changes in the spectroscopic and electrochemical properties, usually obey binary logic and can thus be taken as a basis for information processing at the molecular level. Artificial molecular machines capable of performing logic operations have been reported [46].

1.4
Natural Molecular Motors and Machines

In the last few years, much progress has been made in the elucidation of the moving mechanisms of motor biomolecules, owing to the fact that—in addition to the established physiological and biochemical methods—novel in vitro techniques have been developed which combine optical and mechanical methods to observe the behavior of a single protein.

The most important and best known natural molecular motors are ATP synthase, myosin, and kinesin [47, 48]. ATP synthase is the ubiquitous enzyme that manufactures adenosine triphosphate (ATP) and is a rotary motor powered by a proton gradient [49–52]. The enzymes of the myosin and the kinesin families are linear motors that move along polymer substrates (actin filaments for myosin and microtubules for kinesin), converting the energy of ATP hydrolysis into mechanical work [53, 54]. Motion derives from a mechanochemical cycle, during which the motor protein binds to successive sites along the substrate in such a way as to move forward on average.

Several other biological processes are based on motions, including protein folding and unfolding. Another example is RNA polymerase, which moves along DNA while carrying out transcriptions, thus acting as a molecular motor.

Suitable engineering of natural molecular motors and/or integration of motor proteins within artificial nanodevices has also been obtained [55–57], thereby opening up the possibility of building functional hybrid devices.

For space reasons, we cannot discuss in more detail the basic principles and operation mechanisms of natural molecular motors, and the reader should refer to the cited references.

1.5
Rotaxanes and Catenanes as Artificial Molecular Machines

In principle, molecular motors and machines can be designed starting from several kinds of molecular [58–62] and supramolecular [63–66] systems, including DNA [67–74]. However, most of the systems constructed so far are based on rotaxanes, catenanes, and related structures. Some relevant features of these multicomponent systems will therefore be summarized.

The names of these compounds derive from the Latin words *rota* and *axis* for wheel and axle, and *catena* for chain. Rotaxanes [75] are minimally composed (Fig. 2a) of a dumbbell-shaped molecule surrounded by a macrocyclic compound and terminated by bulky groups (stoppers) that prevent disassembly; catenanes [75] are made of (at least) two interlocked macrocycles or "rings" (Fig. 2b). Rotaxanes and catenanes are appealing systems for the construction of molecular machines because of the peculiar mechanical bonds that keep the molecular components together; hence, relative motions of such molecular components can be easily imagined (Fig. 3).

Important features of these systems derive from non-covalent interactions between components that contain complementary recognition sites. Such interactions, that are also responsible for the efficient template-directed syntheses of rotaxanes and catenanes [76–81], include electron donor-acceptor ability, hydrogen bonding, hydrophobic-hydrophilic character, π-π stacking, coulombic forces and, on the side of the strong interaction limit, metal-ligand bonding. The stability of a specific structure for rotaxanes and

Fig. 2 Schematic representation of a rotaxane (**a**) and a catenane (**b**)

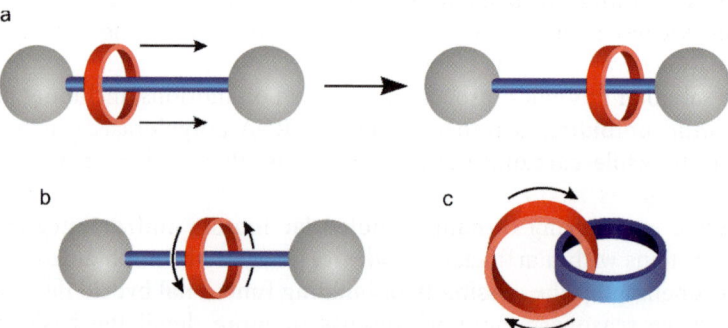

Fig. 3 Some of the intercomponent motions that can be obtained with rotaxanes and catenanes: shuttling (**a**) and ring rotation (**b**, **c**)

catenanes is determined by the intercomponent interactions that can take place. In general, in order to cause controllable mechanical movements, these interactions have to be modulated by means of external stimulation.

Among the most common types of rotaxanes and catenanes are those characterized by (i) charge-transfer (CT) intercomponent interactions between a π-electron acceptor (e.g. a 4,4′-bipyridinium derivative) and a π-electron donor (e.g. a dioxyaromatic unit or a tetrathiafulvalene derivative), and/or (ii) $N^+ - H \cdots O$ intercomponent hydrogen bonding between secondary ammonium functions (e.g. dibenzylammonium ion) and a suitable crown ether (e.g. dibenzo[24]crown-8, DB24C8). The leading group in the synthesis and characterization of these compounds is that of Prof. Fraser Stoddart at the University of California, Los Angeles. All the examples described in this chapter are fruits of a long-lasting collaboration between our group and the Stoddart team.

The charge-transfer interactions between electron donor and electron acceptor units introduce low energy CT excited states (Fig. 4a) which are responsible not only for the presence of broad and weak absorption bands in the visible region, but also for the quenching of the potentially luminescent excited states localized on the molecular components. It should be noted that, when engaged in CT interactions, the electron donor and electron acceptor units become more difficult to oxidize and to reduce, respectively (Fig. 4b). Furthermore, units which are topologically equivalent in an isolated component may not be so when such a component is engaged in non-symmetric interactions with another component. In rotaxanes and catenanes based on CT interactions, mechanical movements can be promoted by destroying such interactions, i.e. reducing the electron acceptor unit(s) or oxidizing the electron donor one(s). This result can be achieved by chemical, electrochemical, or photochemical reactions. In most cases, the CT interactions can be restored by an opposite redox process, which thus promotes a reverse mechanical movement leading back to the original structure.

Contrary to what happens in the case of CT interactions, hydrogen bonds between secondary ammonium centers and crown ethers do not introduce low lying energy levels. Therefore, even if the absorption bands of the molecular components of rotaxanes and catenanes based on this kind of interaction are often perturbed compared with the corresponding absorption bands of the isolated molecular components, no new band is present in the visible region. As far as luminescence is concerned, in the multicomponent architecture each component maintains its potentially luminescent levels, but intercomponent photoinduced energy- and electron-transfer processes can also occur [82, 83]. The electrochemical properties of the separated components are more or less modified when the components are assembled. In these compounds, mechanical movements can be caused by acid-base chemical inputs that strengthen/weaken the hydrogen bonding interactions which

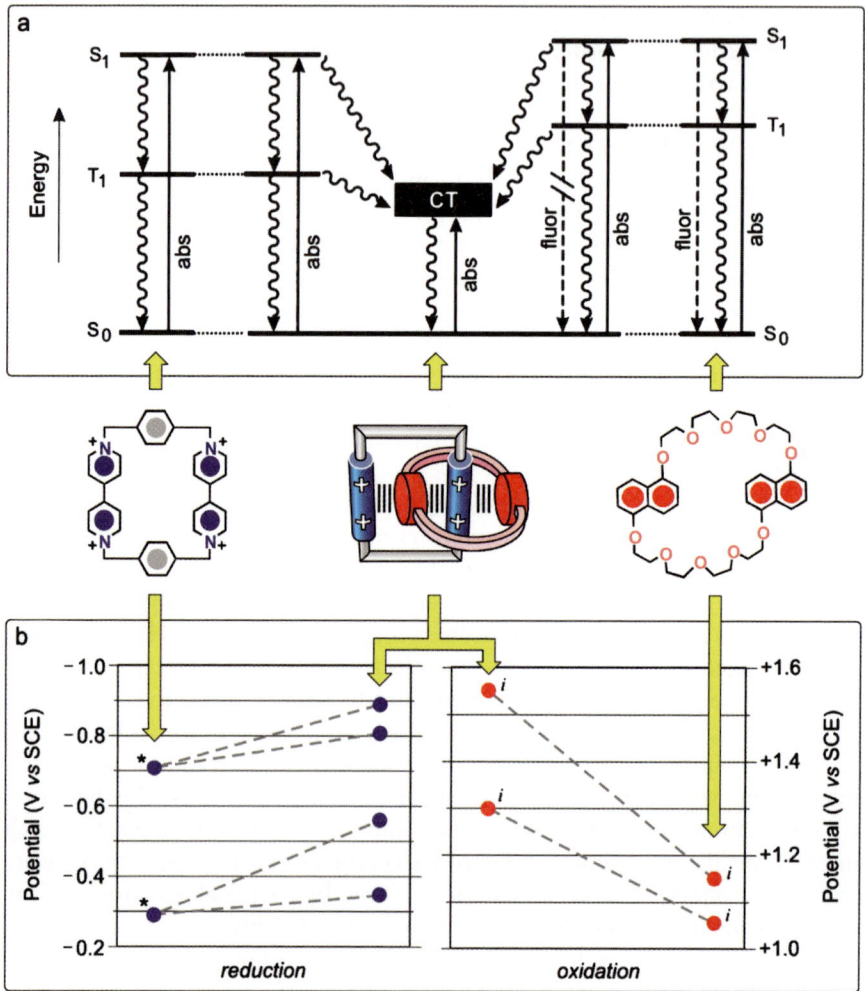

Fig. 4 a Schematic energy level diagram for a catenane based on charge-transfer (CT) interactions and for its separated components. The *wavy lines* indicate non-radiative decay paths of the electronic excited states. **b** Correlations between potential values obtained for the catenane and its separated components (Conditions: acetonitrile solution, room temperature, 5×10^{-4} mol L^{-1}, tetraethylammonium hexafluorophosphate 0.05 mol L^{-1} as the supporting electrolyte, glassy carbon working electrode, scan rate 0.2 V s^{-1}). Processes marked with * involve the exchange of two electrons, while those marked with *i* are irreversible processes

are responsible for assembly and spatial organization. The easiest way to produce an acid-base input is, of course, the addition of suitable chemical species, but photochemical and electrochemical inputs can also be used in principle.

2
Prototypes

2.1
Linear Motions in Rotaxanes

Because of their peculiar structure, at least two interesting molecular motions can be envisaged in rotaxanes, namely (i) translation, i.e. shuttling [84], of the ring along the axle (Fig. 3a); and (ii) rotation of the macrocyclic ring around the axle (Fig. 3b). Hence, rotaxanes are good prototypes for the construction of both rotary and linear molecular motors. Systems of type (i), termed molecular shuttles, constitute indeed the most common implementation of the molecular motor concept with rotaxanes.

If two identical recognition sites for the ring, i.e., "stations", are located within the dumbbell component, the result is a degenerate, conformational equilibrium state in which the ring spontaneously shuttles back and forth along the axle [84]. When the two recognition sites on the dumbbell component are different, a rotaxane can exist as two different equilibrating conformations, the population of which reflects their relative free energy as determined primarily by the strengths of the two different sets of non-covalent bonding

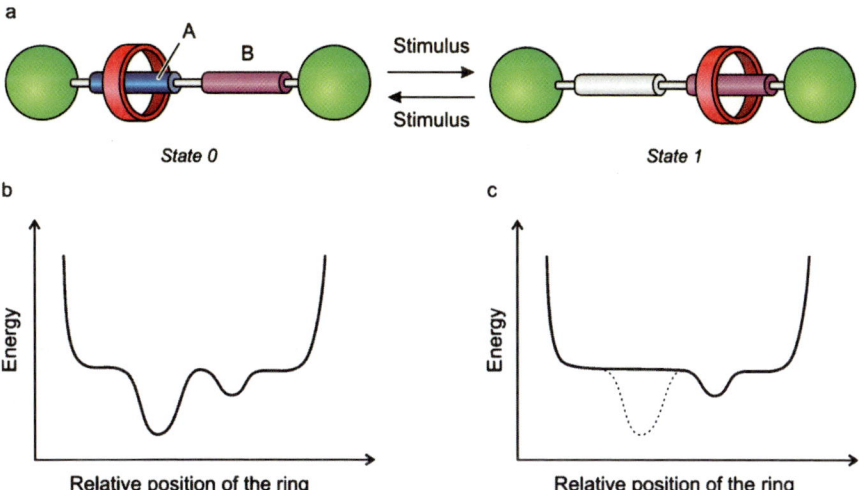

Fig. 5 Schematic representation of a two-station rotaxane and its operation as a controllable molecular shuttle (**a**). The graphs are a simplified representation of the potential energy of the system as a function of the position of the ring relative to the axle before (**b**) and after (**c**) switching off station A. An alternative approach would be to modify station B through an external stimulus in order to make it a stronger recognition site compared to station A

interactions. In the schematic representation shown in Fig. 5, the molecular ring resides preferentially around station A (state 0), until a stimulus is applied that switches off this recognition site. The rotaxane then equilibrates according to the new potential energy landscape, and the molecular ring moves by Brownian motion to the second recognition site (station B, state 1). If station A is switched on again by an opposite stimulus, the original potential energy landscape is restored, and another conformational equilibration occurs through the shuttling of the ring back to station A by Brownian motion. In appropriately designed rotaxanes, the switching process can be controlled by reversible chemical reactions (protonation-deprotonation, reduction-oxidation, isomerization) caused by chemical, electrochemical, or photochemical stimulation.

2.2
An Acid-Base Controlled Molecular Shuttle

The first example of a controllable molecular shuttle was reported in 1994 [85]. Since then, many molecular shuttles relying on chemical [25–28, 86–89], electrochemical [25–28, 90, 91], and photochemical [25–28, 92–94] stimulation have been described in the literature. A chemically driven system with good performance in terms of switching and stability is compound $1H^{3+}$ shown in Fig. 6 [95]. It is made of a dumbbell component containing an ammonium (AMH) and an electron acceptor bipyridinium (BPM) units that can establish hydrogen-bonding and CT interactions, respectively, with the ring component DB24C8, which is a crown ether with electron donor properties. An anthracene moiety is used as a stopper because its absorption, luminescence, and redox properties are useful to monitor the state of the system. Since the $N^+ - H \cdots O$ hydrogen bonding interactions between the macrocyclic ring and the ammonium center are much stronger than the CT interactions of the ring with the bipyridinium unit, the rotaxane exists as only one of the two possible translational isomers (Fig. 6a, state 0). Deprotonation of the ammonium center of $1H^{3+}$ with a base (Fig. 6b) causes 100% displacement of the DB24C8 ring by Brownian motion to the BPM unit (Fig. 6c, state 1); reprotonation of 1^{2+} with an acid (Fig. 6d) directs the ring back on the ammonium center. Such a switching process was investigated in solution by ^1H NMR spectroscopy and by electrochemical and photophysical measurements [95]. Very recently, the kinetics of ring shuttling were also studied in detail by stopped-flow spectroscopic experiments [96]. The full chemical reversibility of the energy supplying acid-base reactions guarantees the reversibility of the mechanical movement, in spite of the formation of waste products. Notice that this rotaxane could be useful for information processing since it is a bistable system and can exhibit a binary logic behavior. It should also be noted that, in the deprotonated rotaxane 1^{2+}, it is possible to displace the ring from the bipyridinium station by destroying the CT interactions through re-

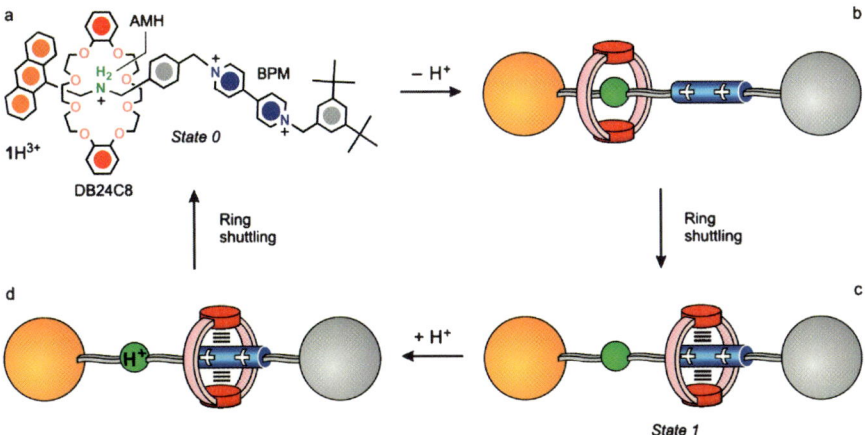

Fig. 6 Schematic representation of the operation of the acid-base controllable molecular shuttle $1H^{3+}$ in solution

duction of the bipyridinium station. Therefore, in this system, mechanical movements can be induced by two different types of stimuli (acid-base and reduction-oxidation).

2.3
A Molecular Elevator

In pursuit of a better fundamental understanding of the nature of multivalency in polytopic hosts and guests, as well as of the intercomponent electronic interactions that occur in complex supramolecular species, we have recently investigated [97] the acid-base controlled (Fig. 7) assembly and disassembly of a triply threaded two-component superbundle. This 1 : 1 adduct consists of a tritopic host **2**, in which three DB24C8 rings are fused together within a triphenylene core, and a trifurcated guest $3H_3^{3+}$ wherein three

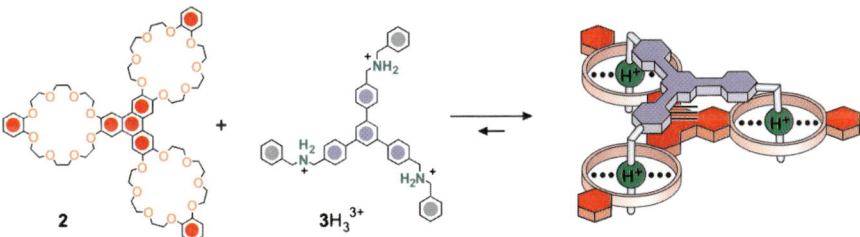

Fig. 7 The self-assembly of the tritopic host **2** and tripod component $3H_3^{3+}$ to afford a triply threaded supramolecular bundle. The adduct can be disassembled and reassembled in solution by addition of base and acid, respectively

dibenzylammonium ions are linked to a central benzenoid core. Fluorescence titration experiments (including Job plots), as well as electrochemical and ^{1}H NMR spectroscopic data in solution, have established the remarkable strength of the superbundle encompassing the triply cooperative binding motif revealed by X-ray crystallography in the solid state. In acetonitrile solution, the dethreading-rethreading of the 1 : 1 adduct can be controlled by addition of base and acid.

By using an incrementally staged strategy, we incorporated the architectural features of the acid-base switchable rotaxane $1H^{3+}/1^{2+}$ (Fig. 6) into those of the trifurcated trication $3H_3^{3+}$ (Fig. 7) and we came up with the de-

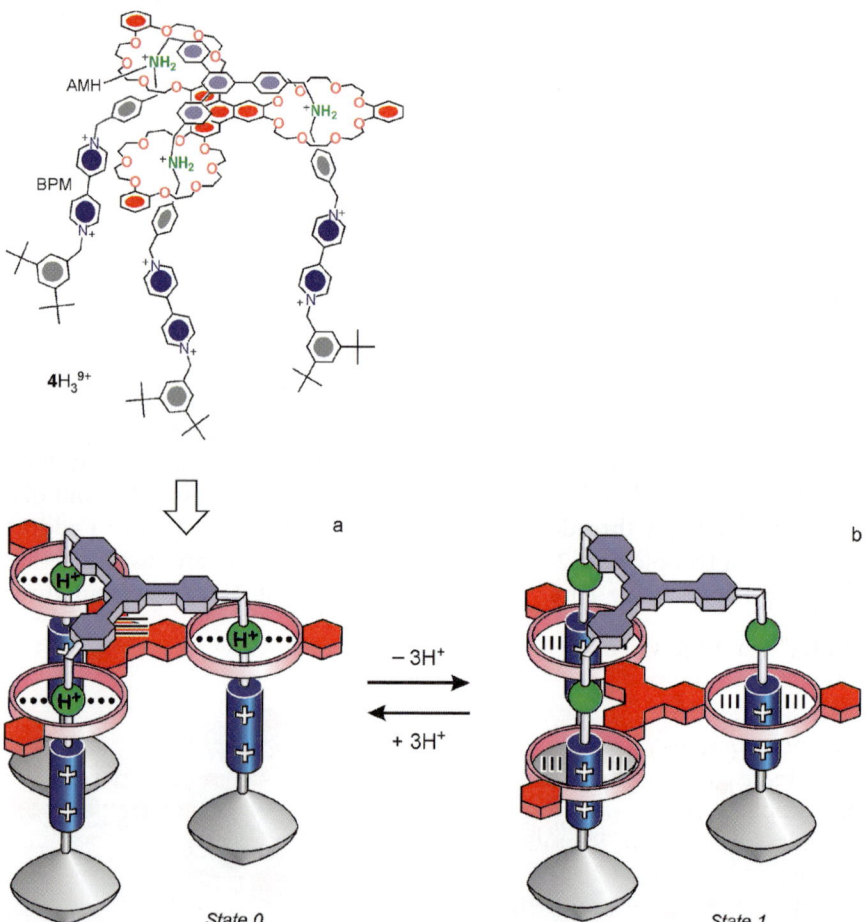

Fig. 8 Chemical formula and operation scheme in solution of the molecular elevator $4H_3^{9+}$, obtained by interlocking the tripod component $5H_3^{3+}$ with the platform-like species 2

sign and construction of a two-component molecular device, $4H_3^{9+}$ (Fig. 8), that behaves like a nanometer-scale elevator [98]. This nanomachine, which is ca. 2.5 nm in height and has a diameter of ca. 3.5 nm, consists of a tripod component $5H_3^{9+}$ containing two different notches—one ammonium center (AMH) and one bipyridinium unit (BPM)—at different levels in each of its three legs. The latter are interlocked by the tritopic host 2, which plays the role of a platform that can be made to stop at the two different levels. The three legs of the tripod carry bulky feet that prevent the loss of the platform. Initially, the platform resides exclusively on the "upper" level [99], i.e., with the three rings surrounding the AMH centers (Fig. 8a, state 0). This preference results from strong $N^+ - H \cdots O$ hydrogen bonding and weak stabilizing π-π stacking forces between the aromatic cores of the platform and tripod components. Upon addition of a strong, non-nucleophilic phosphazene base to an acetonitrile solution of $4H_3^{9+}$, deprotonation of the ammonium center occurs and, as a result, the platform moves to the "lower" level, that is, with the three DB24C8 rings surrounding the bipyridinium units (Fig. 8b, state 1). This structure is stabilized mainly by CT interactions between the electron rich aromatic units of the platform and the electron deficient BPM units of the tripod component. Subsequent addition of acid to 4^{6+} restores the AMH centers, and the platform moves back to the upper level.

The "up and down" elevator-like motion, which corresponds to a quantitative switching and can be repeated many times, can be monitored by spectroscopic techniques (^{1}H NMR, absorption, and fluorescence) and, very conveniently, by electrochemistry. It is well known [100] that the BPM unit exhibits two consecutive reversible one-electron reduction processes at potential values of around -0.4 and -0.8 V versus SCE, and that when such a unit is surrounded by electron donors, such as the dioxybenzene units in DB24C8, its reduction potentials shift to more negative values (see Sect. 1.5). The voltammetric analysis of $4H_3^{9+}$ shows the occurrence of two consecutive reversible reduction processes, each involving the exchange of three electrons (Fig. 9a), at potential values identical to those found for the tripod component $5H_3^{9+}$. This observation indicates that the three BPM units of $4H_3^{9+}$ (and of $5H_3^{9+}$) are equivalent, behave independently from one another, and do not interact with the platform, which consistently resides on the "upper" level. After addition of three equivalents of base, two reversible three-electron reduction processes are still observed in cyclic voltammetric experiments, but they are displaced to more negative values (Fig. 9b). Such a change cannot be due simply to deprotonation of the AMH centers, because the potential values for reduction of the BPM units in the deprotonated tripod component 5^{6+} are identical to those of $5H_3^{9+}$. Hence, the results obtained show that the BPM units in the three legs are surrounded by the electron donor DB24C8 rings of the platform. The changes in reduction potential can be fully reversed by the addition of acid and the cycle can be repeated without any apparent loss of reversibility.

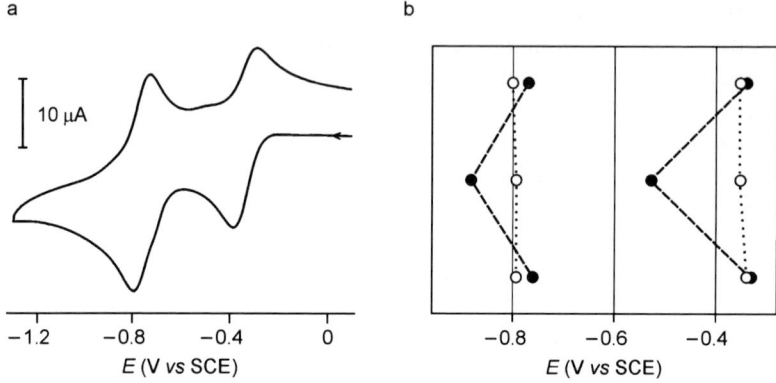

Fig. 9 a Cyclic voltammetry of $4H_3^{9+}$, showing the two reversible three-electron reduction processes of the three bipyridinium units. (Conditions: acetonitrile solution, room temperature, 4×10^{-4} mol L^{-1}, tetrabutylammonium hexafluorophosphate 0.05 mol L^{-1} as the supporting electrolyte, glassy carbon working electrode, scan rate 0.1 V s^{-1}). **b** Correlation diagram of the potential values, in V versus SCE, for the reduction processes of the bipyridinium units of $4H_3^{9+}$ (●) and $5H_3^{9+}$ (○) (*top*), upon addition of a base to afford 4^{6+} and, respectively, 5^{6+} (*middle*), and after reprotonation with an acid—stoichiometric amount with respect to the added base (*bottom*)

The distance traveled by the platform is about 0.7 nm, and from thermodynamic considerations it can be estimated that the elevator movement from the upper to lower level could in principle generate a force of up to 200 pN—one order of magnitude higher than that developed by myosin and kinesin [47, 48].

It should be noted that the acid-base controlled mechanical motion in $4H_3^{9+}$ is associated with interesting structural modifications, such as the opening and closing of a large cavity and the control of the positions and properties of the bipyridinium legs. This behavior can in principle be used to control the uptake and release of a guest molecule, a function of interest for the development of drug delivery systems.

2.4
An Autonomous Light-Powered Molecular Motor

The chemically powered artificial motors described in the previous sections are *not autonomous* since, after the mechanical movement induced by a chemical input, they need another, opposite chemical input to reset, which also implies generation of waste products. However, as illustrated in Sect. 1.3, addition of a reactant (fuel) is not the only means by which energy can be supplied to a chemical system. In fact, Nature shows that, in green plants, the energy needed to sustain the machinery of life is ultimately provided by sunlight. Energy inputs in the form of photons can indeed cause mechanical movements by reversible chemical reactions without formation of waste products. Only a few examples of light driven artificial molecular motors exhibiting autonomous behavior have been reported so far [59, 61, 101].

The design and construction of molecular shuttles powered exclusively by light energy is therefore a fascinating yet challenging subject. On the basis of the experience gained with previous studies on pseudorotaxane model systems [102–105], the rotaxane 6^{6+} (Fig. 10) was specifically designed to achieve photoinduced ring shuttling in solution [106]. This compound is made of the electron donor ring R, and a dumbbell component which contains several units: a Ruthenium(II) polypyridine complex (P) that plays the dual role of a light-fueled motor [107] and a stopper, a p-terphenyl-type rigid spacer (S), a 4,4'-bipyridinium unit (A_1) and a 3,3'-dimethyl-4,4'-bipyridinium unit (A_2) as electron accepting stations, and a tetraarylmethane group as the second stopper (T). The stable translational isomer of rotaxane 6^{6+} is the one in which the R component encircles the A_1 unit, in keeping with the fact that this station is a better electron acceptor than the other one.

The strategy devised in order to obtain the photoinduced abacus-like movement of the R macrocycle between the two stations A_1 and A_2, illustrated in Fig. 11, is based on the following four operations:

(a) *Destabilization of the stable translational isomer*: light excitation of the photoactive unit P (process 1) is followed by the transfer of an electron from the excited state to the A_1 station, which is encircled by the ring R (process 2), with the consequent "deactivation" of this station; such

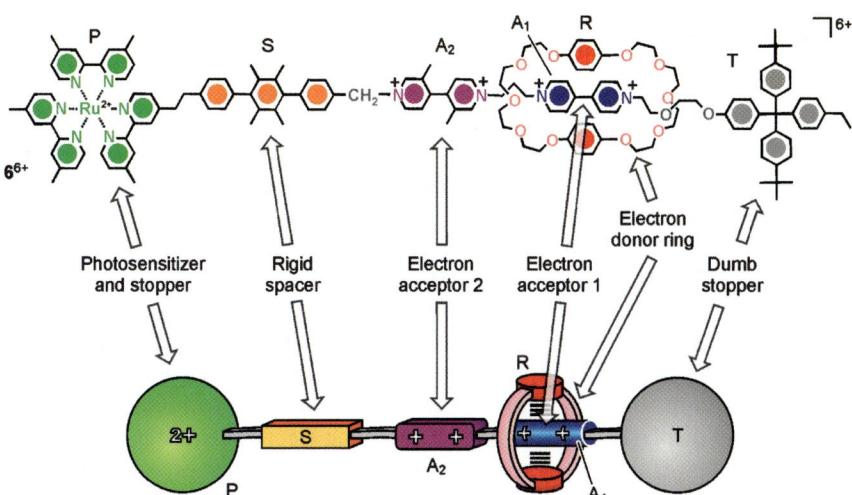

Fig. 10 Chemical formula and cartoon representation of the rotaxane 6^{6+}, showing its modular structure

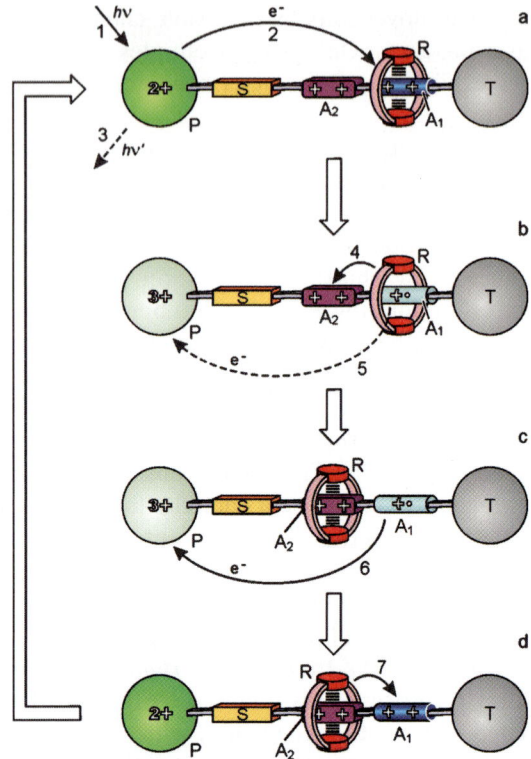

Fig. 11 Schematic representation of the operation of rotaxane 6^{6+} as an autonomous "four stroke" linear nanomotor powered by light

a photoinduced electron-transfer process has to compete with the intrinsic decay of the P excited state (process 3).
(b) *Ring displacement*: the ring moves by Brownian motion (process 4) from the reduced A_1 station to A_2, a step that has to compete with the back electron-transfer process from reduced A_1 (still encircled by R) to the oxidized P unit (process 5).
(c) *Electronic reset*: a back electron-transfer process from the "free" reduced A_1 station to the oxidized P unit (process 6) restores the electron acceptor power to the A_1 station.
(d) *Nuclear reset*: as a consequence of the electronic reset, back movement of the ring by Brownian motion from A_2 to A_1 takes place (process 7).

The crucial point for such a mechanism is indeed the favorable competition between ring displacement (process 4) and back electron transfer (process 5). The rate constants of the relevant electron-transfer processes were measured in acetonitrile solution by laser flash photolysis [106]; however, direct time-resolved observation of the ring displacement proved to be quite elusive. Very recently, we performed laser flash photolysis experiments in the presence of an electron relay in order to slow down the back electron-transfer process and facilitate the observation of ring displacement [108]. We observed that the transient absorption spectrum of the photogenerated one-electron reduced A_1 unit in the rotaxane changes slightly with time, while it does not change at all for the dumbbell component. Such changes have been attributed to the motion of the ring R, which immediately after light excitation surrounds the reduced A_1 unit and subsequently moves away from it to encircle the A_2 station.

These investigations revealed that in acetonitrile at room temperature the ring shuttling rate is one order of magnitude slower than the back electron transfer. Hence, the absorption of a visible photon can cause the occurrence of a forward and back ring movement (i.e. a full cycle) without generation of waste products, but with a low (around 2%) quantum efficiency. The low efficiency is compensated by the fact that the operation of the system relies *exclusively* on intramolecular processes. Therefore, this artificial molecular motor does not need the assistance of external species and, in principle, it can work at the single-molecule level. In other words, 6^{6+}, performs as a "four stroke" autonomous artificial linear motor working by an *intramolecular* mechanism powered by *visible* light.

2.5
Systems Based on Catenanes

Catenanes are chemical compounds consisting minimally of two interlocked macrocycles [75]. When one of the two rings carries two different recognition sites, then the opportunity exists to control the dynamic processes in a manner reminiscent of the controllable molecular shuttles. By switching *off* and *on* again the recognition properties of one of the two recognition sites of the non-symmetric ring by means of external energy stimuli, it is indeed possible to induce conformational changes that can be viewed as the rotation of the non-symmetric ring.

An example of such a behavior is offered by the catenane 7^{4+} shown in Fig. 12 [109, 110]. This compound is made of a symmetric tetracationic ring containing two electron acceptor bipyridinium units and a non-symmetric ring comprising two different electron donor units, namely a tetrathiafulvalene (TTF) group and a 1,5-dioxynaphthalene (DON) unit. Since the TTF unit is a better electron donor than the DON one, as witnessed by the potentials values for their oxidation, the thermodynamically stable conformation of the catenane is that in which the symmetric ring encircles the TTF unit of the non-symmetric one (Fig. 12a, state 0). On electrochemical oxidation in solution, the TTF unit loses its electron donor power and acquires a positive charge (Fig. 12b). As a consequence it is expelled from the cavity of the tetracationic ring and is replaced by the neutral DON unit (Fig. 12c, state 1). At this stage, subsequent reduction of the oxidized TTF unit restores its electron donor ability and the

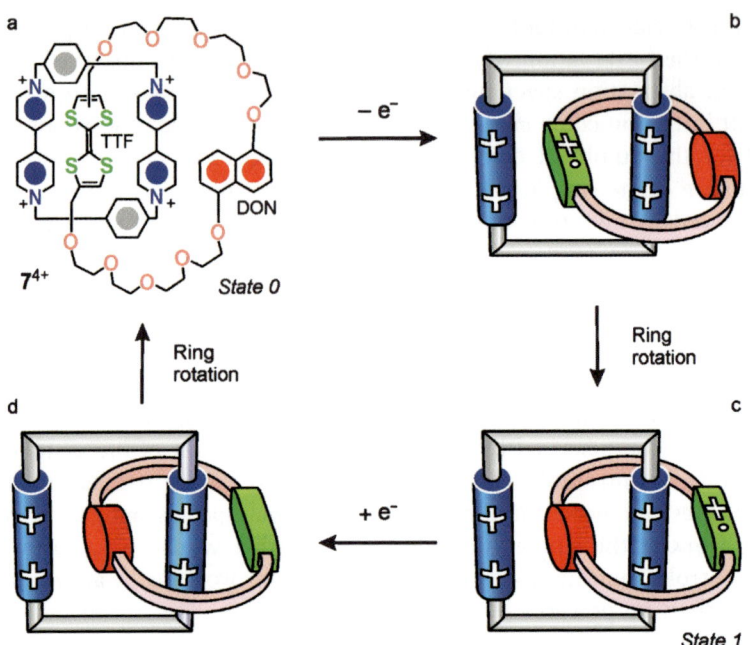

Fig. 12 Redox controlled ring rotation in solution for catenane 7^{4+}, which contains a non-symmetric ring

system goes back to its original conformation. A variety of techniques, including cyclic voltammetry, were employed to characterize the system. The catenane 7^{4+} was also incorporated in a solid state device that could be used for random access memory (RAM) storage [111, 112]. Additionally, this compound could be employed for the construction of electrochromic systems, because its various redox states are characterized by different colors [109, 110, 113].

It should be pointed out that in the catenane system described above, repeated switching between the two states does not need to occur through a full rotation. In fact, because of the intrinsic symmetry of the system, both the movement from state 0 to state 1 and that from state 1 to state 0 can take place, with equal probabilities, along a clockwise or anticlockwise direction. A full (360°) rotation movement, which would be much more interesting from a mechanical viewpoint, can only occur in ratchet-type systems, i.e. in the presence of asymmetry elements which can be structural or functional in nature [40, 114]. This idea was recently implemented with a carefully designed catenane by relying on a sequence of photochemical, chemical, and thermally activated processes [115] and employing ^{1}H NMR spectroscopy to characterize the system.

By an appropriate choice of the functional units that are incorporated in the catenane components, more complex functions can be obtained. An example is represented by catenane $8H^{5+}$ (Fig. 13), composed of a symmetric crown ether ring and a cyclophane ring containing two bipyridinium (BPM) and one ammonium (AMH) recognition sites [116]. The absorption spectra and electrochemical properties show that the crown ether ring surrounds a BPM unit of the other ring both in $8H^{5+}$ (Fig. 13a) and in its deprotonated form 8^{4+} (Fig. 13b), indicating that deprotonation-protonation of the AMH unit does not cause any displacement of the crown ether ring (state 0). Electrochemical measurements show that, after one-electron reduction of both the BPM units of $8H^{5+}$, the crown ether ring is displaced on the AMH function (Fig. 13c, state 1), which means that an electrochemically induced conformational switching does occur. Furthermore, upon deprotonation of the two-electron reduced form $8H^{3+}$ (Fig. 13d), the macrocyclic polyether moves to one of the monoreduced BPM units (state 0). Therefore, in order to achieve the motion of the crown ether ring in the deprotonated catenane 8^{4+}, it is necessary *both* to reduce (switch off) the BPM units *and* protonate (switch on) the amine function.

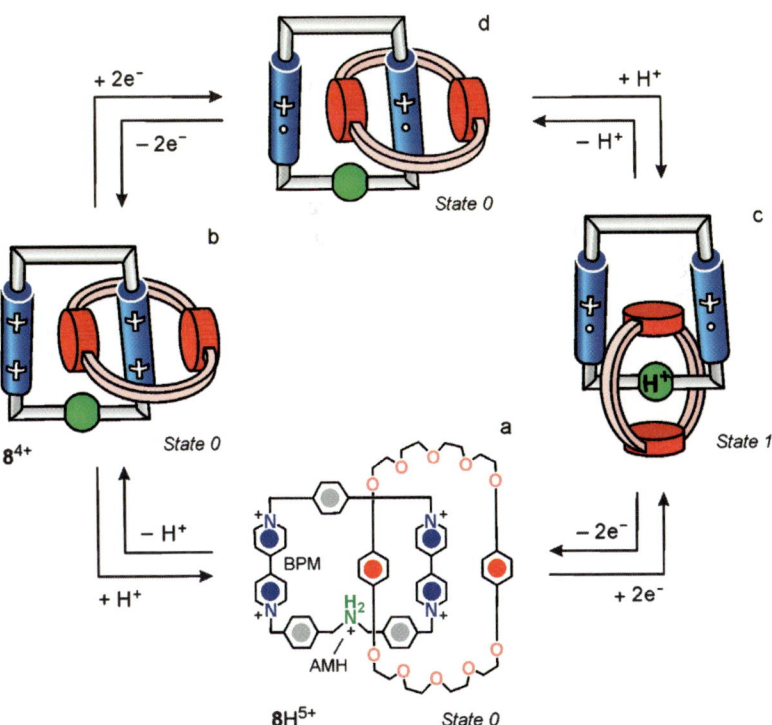

Fig. 13 Switching processes of catenane $8H^{5+}$ in solution. Starting from the deprotonated catenane 8^{4+}, the position of the crown ether ring switches under acid-base and redox inputs according to AND logic

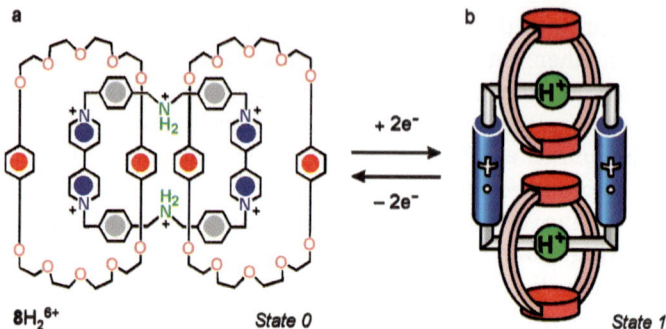

Fig. 14 Redox controlled movements of the ring components in a catenane composed of three interlocked macrocycles. These motions are obtained upon reduction-oxidation of the bipyridinium units of the cyclophane ring in solution

The mechanical motion in such a catenane takes place according to an AND logic [46], a function associated with two energy inputs of different nature.

Controlled rotation of the molecular rings has been achieved also in catenanes composed of three interlocked macrocycles. For example, catenane $9H_2^{6+}$ (Fig. 14) is made up of two identical crown ether rings interlocked with a cyclophane ring containing two BPM and two AMH units [116]. Because of the type of crown ethers used, the stable conformation of $9H_2^{6+}$ is that where the two crown ether rings surround the BPM units (Fig. 14a, state 0). Upon addition of one electron in each of the BPM units, the two macrocycles move on the AMH stations (Fig. 14b, state 1), and move back to the original position when the bipyridinium units are reoxidized. A clever—albeit complex—way to obtain a unidirectional full rotation in a catenane having the same topology of $9H_2^{6+}$ was devised by another research group [117]. Other examples of molecular motors based on catenanes can be found in the literature [25–28, 118].

3
Conclusion and Perspectives

The results described here show that, by taking advantage of careful incremental design strategies, of the tools of modern synthetic chemistry, of the paradigms of supramolecular chemistry, as well as inspiration from natural systems, it is possible to produce compounds capable of performing non-trivial mechanical movements and exercising a variety of different functions upon external stimulation.

In the previously mentioned address to the American Physical Society [2, 3], R. P. Feynman concluded his reflection on the idea of constructing molecular machines as follows: *"What would be the utility of such machines? Who knows? I cannot see exactly what would happen, but I can hardly doubt that when we have some control of the rearrangement of things on a molecular scale we will get an enormously greater range of possible properties that substances can have, and of different things we can do"*. This sentence, pronounced in 1959, is still an appropriate comment to the work described in this chapter. The results achieved sowed the seeds for future developments, which are under investigation in our laboratory: (i) the design and construction of more sophisticated artificial molecular motors and machines; (ii) the use of such systems to do tasks such as molecular-level transportation, catalysis, and mechanical gating of molecular channels; and (iii) the possibility of exploiting their logic behavior for information processing at the molecular level and, in the long run, for the construction of chemical computers.

It should also be noted that the majority of the artificial molecular motors developed so far operate in solution, that is, in an incoherent fashion and without control of spatial positioning. The studies in solution of complicated chemical systems such as molecular motors and machines are indeed of fundamental importance to understand their operation mechanisms; moreover, for some uses (e.g., drug delivery) molecular machines will have to work in liquid solution. In this regard, it should be recalled that motor proteins operate in—or at least in contact with —an aqueous solution. However, it seems reasonable that, before artificial molecular motors and machines can find applications in many fields of technology, they have to be interfaced with the macroscopic world by ordering them in some way. The next generation of molecular machines and motors will need to be organized at interfaces, deposited on surfaces, or immobilized into membranes or porous materials [111–113, 119–126] so that they can behave coherently and can be addressed in space. Indeed, the preparation of modified electrodes represents one of the most promising ways to achieve this goal.

Apart from more or less futuristic applications, the extension of the concepts of motor and machine to the molecular level is of interest not only for the development of nanotechnology, but also for the growth of basic research. Looking at molecular and supramolecular species from the viewpoint of functions with references to devices of the macroscopic world is indeed a very interesting exercise which introduces novel concepts into Chemistry as a scientific discipline.

Acknowledgements We would like to thank Prof. J. Fraser Stoddart and his group for a long lasting and fruitful collaboration. Financial support from the EU (STREP "Biomach" NMP2-CT-2003-505487), Ministero dell'Istruzione, dell'Università e della Ricerca (PRIN "Supramolecular Devices" and FIRB RBNE019H9K), and Università di Bologna (Funds for Selected Research Topics) is gratefully acknowledged.

References

1. Keyes RW (2001) Proc IEEE 89:227, see also the International Technology Roadmap for Semiconductors (ITRS), 2004 Edition, available at http://public.itrs.net
2. Feynman RP (1960) Eng Sci 23:22
3. Feynman RP (1960) Saturday Rev 43:45; see also: http://www.feynmanonline.com
4. Drexler KE (1986) Engines of Creation—The Coming Era of Nanotechnology. Anchor Press, New York
5. Drexler KE (1992) Nanosystems. Molecular Machinery, Manufacturing, and Computation. Wiley, New York
6. Smalley RE (2001) Sci Am 285:76
7. Joachim C, Launay JP (1984) Nouv J Chem 8:723
8. Balzani V, Moggi L, Scandola F (1987) In: Balzani V (ed) Supramolecular Photochemistry. Reidel, Dordrecht, p 1
9. Lehn JM (1990) Angew Chem Int Ed Engl 29:1304
10. Aviram A, Ratner MA (1974) Chem Phys Lett 29:277
11. Carter FL (ed) (1982) Molecular Electronic Devices. Dekker, New York
12. Metzger RM (2003) Chem Rev 103:3803 and references therein
13. Lehn JM (1995) Supramolecular Chemistry—Concepts and Perspectives. VCH, Weinheim
14. Lehn JM, Atwood JL, Davies JED, MacNicol DD, Vögtle F (eds) (1996) Comprehensive Supramolecular Chemistry. Pergamon, Oxford
15. Steed JW, Atwood JL (2000) Supramolecular Chemistry. Wiley, New York
16. Atwood JL, Steed JW (eds) (2004) Encyclopedia of Supramolecular Chemistry. Dekker, New York
17. Rigler R, Orrit M, Talence I, Basché T (2001) Single Molecule Spectroscopy. Springer, Berlin Heidelberg New York
18. Moerner WE (2002) J Phys Chem B 106:910
19. Zander C, Enderlein J, Keller RA (2002) Single Molecule Detection in Solution. Wiley-VCH, Weinheim
20. Gimzewski JK, Joachim C (1999) Science 283:1683
21. Hla SW, Meyer G, Rieder KH (2001) Chem Phys Chem 2:361
22. Samorì B, Zuccheri G, Baschieri P (2005) Chem Phys Chem 6:29
23. See e.g.: Christ T, Kulzer F, Bordat P, Basché T (2001) Angew Chem Int Ed 40:4192
24. Goodsell DS (2004) Bionanotechnology—Lessons from Nature. Wiley, New York
25. Balzani V, Credi A, Raymo FM, Stoddart JF (2000) Angew Chem Int Ed 39:3348
26. (2001) Acc Chem Res 34, no 6. Special Issue on Molecular Machines
27. (2001) Struct Bond 99. Special Volume on Molecular Machines and Motors
28. Balzani V, Credi A, Venturi M (2003) Molecular Devices and Machines—A Journey into the Nano World. Wiley-VCH, Weinheim
29. Sauvage JP (2005) Chem Comm 1507
30. Akkerman OS, Coops J (1967) Rec Trav Chim Pays-Bas 86:755
31. Cozzi F, Guenzi A, Johnson CA, Mislow K, Hounshell WD, Blount JF (1981) J Am Chem Soc 103:957
32. Bedard TC, Moore JS (1995) J Am Chem Soc 117:10662
33. Dominguez Z, Khuong TAV, Dang H, Sanrame CN, Nunez JE, Garcia-Garibay MA (2003) J Am Chem Soc 125:8827
34. Shima T, Hampel F, Gladysz JA (2004) Angew Chem Int Ed 43:5537
35. Khuong TAV, Zepeda G, Sanrame CN, Dang H, Bartbeger MD, Houk KN, Garcia-Garibay MA (2004) J Am Chem Soc 126:14778

36. Kottas GS, Clarke LI, Horinek D, Michl J (2005) Chem Rev 105:1281
37. Astumian RD, Hänggi P (2002) Phys Today 55:33
38. Parisi G (2005) Nature 433:221
39. Astumian RD (2005) Proc Natl Acad Sci USA 102:1843
40. Ballardini R; Balzani V, Credi A, Gandolfi MT, Venturi M (2001) Acc Chem Res 34:445
41. Balzani V, Scandola F (1991) Supramolecular Photochemistry. Horwood, Chichester
42. Armaroli N (2003) Photochem Photobiol Sci 2:73
43. Balzani V (2003) Photochem Photobiol Sci 2:479
44. Kaifer AE, Gómez-Kaifer M (1999) Supramolecular Electrochemistry. Wiley-VCH, Weinheim
45. Marcaccio M, Paolucci F, Roffia S (2004) In: Pombeiro AJL, Amatore C (eds) Trends in Molecular Electrochemistry. Dekker, New York, p 223
46. Balzani V, Credi A, Venturi M (2003) ChemPhysChem 4:49
47. Schliwa M (ed) (2003) Molecular Motors. Wiley-VCH, Weinheim
48. Schliwa M, Woehlke G (2003) Nature 422:759
49. Boyer PD (1993) Biochim Biophys Acta 1140:215
50. Boyer PD (1998) Angew Chem Int Ed 37:2296
51. Walker JE (1998) Angew Chem Int Ed 37:2308
52. Stock D, Leslie AGW, Walker JE (1999) Science 286:700
53. Vale RD, Milligan RA (2000) Science 288:88
54. Frey E (2002) ChemPhysChem 3:270
55. Steinberg-Yfrach G, Rigaud JL, Durantini EN, Moore AL, Gust D, Moore TA (1998) Nature 392:479
56. Soong RK, Bachand GD, Neves HP, Olkhovets AG, Craighead HG, Montemagno CD (2000) Science 290:1555
57. Hess H, Bachand GD, Vogel V (2004) Chem Eur J 10:2110
58. Kelly TR, De Silva H, Silva RA (1999) Nature 401:150
59. Koumura N, Zijlstra RWJ, van Delden RA, Harada N, Feringa BL (1999) Nature 401:152
60. Kelly TR, Silva RA, De Silva H, Jasmin S, Zhao YJ (2000) J Am Chem Soc 12:6935
61. van Delden RA, Koumura N, Schoevaars A, Meetsma A, Feringa BL (2003) Org Biomol Chem 1:33
62. Zheng XL, Mulcahy ME, Horinek D, Galeotti F, Magnera TF, Michl J (2004) J Am Chem Soc 126:4540
63. Shinkai S, Ikeda M, Sugasaki A, Takeuchi M (2001) Acc Chem Res 34:494
64. Gray M, Cuello AO, Cooke G, Rotello VM (2003) J Am Chem Soc 125:7882
65. Moon K, Grindstaff J, Sobransingh D, Kaifer AE (2004) Angew Chem Int Ed 43:5496
66. Jeon WS, Kim E, Ko YH, Hwang IH, Lee JW, Kim SY, Kim HJ, Kim K (2005) Angew Chem Int Ed 44:87
67. Mao C, Sun W, Shen Z, Seeman NC (1999) Nature 397:144
68. Yurke B, Turberfield AJ, Mills AP Jr, Simmel FC, Neumann JL (2000) Nature 406:605
69. Yan H, Zhang X, Shen Z, Seeman NC (2002) Nature 415:82
70. Li JJ, Tan W (2002) Nano Lett 2:315
71. Chen Y, Wang M, Mao C (2004) Angew Chem Int Ed 43:3554
72. Chen Y, Mao C (2004) J Am Chem Soc 126:8626
73. Yin P, Yan H, Daniell XG, Turberfield AJ, Reif JH (2004) Angew Chem Int Ed 43:4906
74. Sherman WB, Seeman NC (2004) Nano Lett 4:1203
75. Sauvage JP, Dietrich-Buchecker C (eds) (1999) Molecular Catenanes, Rotaxanes and Knots. Wiley-VCH, Weinheim

76. Chambron JC, Dietrich-Buchecker CO, Sauvage JP (1993) Top Curr Chem 165:131
77. Gibson HW, Bheda MC, Engen PT (1994) Prog Polym Sci 19:843
78. Amabilino DB, Stoddart JF (1995) Chem Rev 95:2725
79. Johnston AG, Leigh DA, Pritchard RJ, Degan MD (1995) Angew Chem Int Ed Engl 34:1209
80. Jäger R, Vögtle F (1997) Angew Chem Int Ed 36:930
81. Fujita M (1999) Acc Chem Res 32:53
82. Ishow E, Credi A, Balzani V, Spadola F, Mandolini L (1999) Chem Eur J 5:984
83. Ballardini R, Balzani V, Clemente-Léon M, Credi A, Gandolfi MT, Ishow E, Perkins J, Stoddart JF, Tseng HR, Wenger S (2002) J Am Chem Soc 124:12786
84. Anelli PL, Spencer N, Stoddart JF (1991) J Am Chem Soc 113:5131
85. Bissell A, Córdova E, Kaifer AE, Stoddart JF (1994) Nature 369:133
86. Elizarov AM, Chiu SH, Stoddart JF (2002) J Org Chem 67:9175
87. Tseng HR, Vignon SA, Stoddart JF (2003) Angew Chem Int Ed 42:1491
88. Keaveney CM, Leigh DA (2004) Angew Chem Int Ed 43:1222
89. Leigh DA, Perez EM (2004) Chem Comm 2262
90. Tseng HR, Vignon SA, Celestre PC, Perkins J, Jeppesen JO, Di Fabio A, Ballardini R, Gandolfi MT, Venturi M, Balzani V, Stoddart JF (2004) Chem Eur J 10:155
91. Altieri A, Gatti FG, Kay ER, Leigh DA, Martel D, Paolucci F, Slawin AMZ, Wong JKY (2003) J Am Chem Soc 125:8644
92. Willner I, Pardo-Yssar V, Katz E, Ranjit KT (2001) J Electroanal Chem 497:172
93. Wang QC, Qu DH, Ren J, Chen KC, Tian H (2004) Angew Chem Int Ed 43:2661
94. Abraham W, Grubert L, Grummt UW, Buck K (2004) Chem Eur J 10:3562
95. Ashton PR, Ballardini R, Balzani V, Baxter I, Credi A, Fyfe MCT, Gandolfi MT, Gomez-Lopez M, Martinez-Diaz MV, Piersanti A, Spencer N, Stoddart JF, Venturi M, White AJP, Williams DJ (1998) J Am Chem Soc 120:11932
96. Garaudée S, Credi A, Silvi S, Venturi M, Stoddart JF (2005) ChemPhysChem 6:2145
97. Balzani V, Clemente-Leon M, Credi A, Lowe JN, Badjic JD, Stoddart JF, Williams DJ (2003) Chem Eur J 9:5348
98. Badjic JD, Balzani V, Credi A, Silvi S, Stoddart JF (2004) Science 303:1845
99. The molecular elevator operates in solution, i.e. with no control of the orientation of the molecules relative to a fixed reference system. Therefore, in the present context the words "upper" and "lower" are used only for descriptive purposes
100. Monk PMS (1998) The Viologens—Physicochemical Properties, Synthesis and Application of the Salts of 4,4'-Bipyridine. Wiley, Chichester
101. Brouwer AM, Frochot C, Gatti FG, Leigh DA, Mottier L, Paolucci F, Roffia S, Wurpel GWH (2001) Science 291:2124
102. Ballardini R, Balzani V, Gandolfi MT, Prodi L, Venturi M, Philp D, Ricketts HG, Stoddart JF (1993) Angew Chem Int Ed Engl 32:1301
103. Ashton PR, Ballardini R, Balzani V, Boyd SE, Credi A, Gandolfi MT, Gomez-Lopez M, Iqbal S, Philp D, Preece JA, Prodi L, Ricketts HG, Stoddart JF, Tolley MS, Venturi M, White AJP, Williams DJ (1999) Chem Eur J 3:152
104. Ashton PR, Ballardini R, Balzani V, Constable EC, Credi A, Kocian O, Langford SJ, Preece JA, Prodi L, Schofield ER, Spencer N, Stoddart JF, Wenger S (1998) Chem Eur J 4:2413
105. Ashton PR, Balzani V, Kocian O, Prodi L, Spencer N, Stoddart JF (1998) J Am Chem Soc 120:11190
106. Ashton PR, Ballardini R, Balzani V, Credi A, Dress R, Ishow E, Kleverlaan CJ, Kocian O, Preece JA, Spencer N, Stoddart JF, Venturi M, Wenger S (2000) Chem Eur J 6:3558

107. Ballardini R, Balzani V, Credi A, Gandolfi MT, Venturi M (2001) Int J Photoenergy 3:63
108. Balzani V, Clemente-León M, Credi A, Ferrer B, Venturi M, Flood AH, Stoddart JF (submitted)
109. Asakawa M, Ashton PR, Balzani V, Credi A, Hamers C, Mattersteig G, Montalti M, Shipway AN, Spencer N, Stoddart JF, Tolley MS, Venturi M, White AJP, Williams DJ (1998) Angew Chem Int Ed 37:333
110. Balzani V, Credi A, Mattersteig G, Matthews OA, Raymo FM, Stoddart JF, Venturi M, White AJP, Williams DJ (2000) J Org Chem 65:1924
111. Collier CP, Mattersteig G, Wong EW, Luo Y, Beverly K, Sampaio J, Raymo FM, Stoddart JF, Heath JR (2000) Science 289:1172
112. Luo Y, Collier CP, Jeppesen JO, Nielsen KA, Delonno E, Ho G, Perkins J, Tseng HR, Yamamoto T, Stoddart JF, Heath JR (2002) ChemPhysChem 3:519
113. Steuerman DW, Tseng HR, Peters AJ, Flood AH, Jeppesen JO, Nielsen KA, Stoddart JF, Heath JR (2004) Angew Chem Int Ed 43:6486
114. Ref [28], ch 16
115. Hernández JV, Kay ER, Leigh DA (2004) Science 306:1532
116. Ashton PR, Baldoni V, Balzani V, Credi A, Hoffmann HDA, Martinez-Diaz MV, Raymo FM, Stoddart JF, Venturi M (2001) Chem Eur J 7:3482
117. Leigh DA, Wong JKY, Dehez F, Zerbetto F (2003) Nature 424:174
118. Mobian P, Kern JM, Sauvage JP (2004) Angew Chem Int Ed 43:2392
119. Cavallini M, Biscarini F, Leon S, Zerbetto F, Bottari G, Leigh DA (2003) Science 299:531
120. Long B, Nikitin K, Fitzmaurice D (2003) J Am Chem Soc 125:5152
121. Álvaro M, Ferrer B, García H, Palomares EJ, Balzani V, Credi A, Venturi M, Stoddart JF, Wenger S (2003) J Phys Chem B 107:14319
122. Huang TJ, Tseng HR, Sha L, Lu WX, Brough B, Flood AH, Yu BD, Celestre PC, Chang JP, Stoddart JF, Ho CM (2004) Nano Lett 4:2065
123. Hernandez R, Tseng HR, Wong JW, Stoddart JF, Zink JI (2004) J Am Chem Soc 126:3370
124. Katz E, Lioubashevsky O, Willner I (2004) J Am Chem Soc 126:15520
125. Cecchet F, Rudolf P, Rapino S, Margotti M, Paolucci F, Baggerman J, Brouwer AM, Kay ER, Wong JKY, Leigh DA (2004) J Phys Chem B 108:15192
126. Flood AH, Peters AJ, Vignon SA, Steuerman DW, Tseng HR, Kang S, Heath JR, Stoddart JF (2004) Chem Eur J 10:6558

Transition-Metal-Complexed Catenanes and Rotaxanes in Motion: Towards Molecular Machines

Jean-Paul Collin · Valérie Heitz · Jean-Pierre Sauvage (✉)

Laboratoire de Chimie Organo-Minérale, UMR 7513 du CNRS, Institut Le Bel,
Université Louis Pasteur, 4 rue Blaise Pascal, 67000 Strasbourg Cedex, France
sauvage@chimie.u-strasbg.fr

1	Introduction	30
2	**Electrochemically Driven Molecular Machines Based on the Copper(II/I) Couple**	32
2.1	The First Catenane Whose Motions are Controlled by Oxidizing or Reducing the Central Copper Atom	32
2.2	Copper-Complexed [2]Catenane in Motion with Three Distinct Geometries	35
2.3	A Molecular Shuttle: Gliding of a Ring on its Axle Under the Action of an Electrochemical Signal	38
2.4	Transition-Metal-Containing Rotaxane in Motion: Electrochemically Induced Pirouetting of the Ring on a Threaded Dumbbell	41
2.5	Fast-Moving Electrochemically Driven Machine Based on a Pirouetting Copper-Complexed Rotaxane	44
3	**Molecular Motions Triggered by an External Chemical Signal**	46
3.1	Complete Rearrangement of a Multi-porphyrinic Rotaxane by Metallation–Demetallation of the Central Coordination Site	46
3.2	Use of a Chemical Reaction to Induce the Contraction/Stretching Process of a Muscle-Like Rotaxane Dimer	48
4	**Ruthenium(II)-Complexed Catenanes and Rotaxanes: Light-Driven Molecular Machines Based on Dissociative Excited States**	50
4.1	Use of Dissociative Excited States to Set Ru(II)-Complexed Molecular Machines in Motion: Principle	50
4.2	Templated Synthesis of a Pseudo-Rotaxane with a $[Ru(diimine)_3]^{2+}$ Core and its Light-Driven Unthreading Reaction	51
4.3	Construction of a [2]Catenane Around a $Ru(diimine)_3^{2+}$ Complex Used as Template	54
4.4	Photoinduced Decoordination and Thermal Recoordination of a Ring in a Ruthenium(II)-Containing [2]Catenane	57
5	Conclusion and Prospective	58
	References	59

Abstract In the course of the last decade, many dynamic molecular systems, for which the movements are controlled from the outside, have been elaborated. These compounds are generally referred to as "molecular machines". Transition-metal-containing catenanes and rotaxanes are ideally suited to build such systems. In the present review article, we will discuss a few examples of molecular machines elaborated and studied in Strasbourg. In the first section we will discuss an *electrochemically driven* system, consisting of copper-complexed catenanes and rotaxanes of various types, including a fast-moving pirouetting rotaxane. The second part will be devoted to *chemically driven* dynamic molecular systems. A porphyrin-stoppered rotaxane, able to undergo a pirouetting motion by metallation or demetallation of the central coordination site, and a linear rotaxane dimer whose behavior is reminiscent of muscles, will be discussed. In both compounds, the mobile component(s) will be set in motion by modifying the central coordination of the molecular assembly. In the rest of this review article, we will mostly focus on *light-driven* machines, consisting of ruthenium(II)-complexed rotaxanes or catenanes. For these latter systems, the synthetic approach is based on the template effect of an octahedral ruthenium(II) center. Two 1,10-phenanthroline ligands are incorporated in an axis or in a ring, affording the precursor to the rotaxane or the catenane, respectively. Ru(diimine)$_3^{2+}$ complexes display the universally used ^{3}MLCT (metal-to-ligand charge transfer) excited state and, another interesting excited state, the ^{3}LF (ligand field) state, which is strongly dissociative. By taking advantage of this latter state, it has been possible to propose a new family of molecular machines, which are set in motion by populating the dissociative ^{3}LF state, thus leading to ligand exchange in the coordination sphere of the ruthenium(II) center.

Keywords Catenane · Molecular machine · Rotaxane · Transition metal

1
Introduction

Molecules are dynamic species in solution and even in the solid state. Processes such as the chair/boat equilibrium of cyclohexane, the nitrogen inversion in amines, or the rotation about a C–C bond in biphenyl derivatives have been studied for decades and are now in all the organic chemistry textbooks. Very different and much more recent is the elaboration and the study of compounds for which the motions can be triggered and controlled at will by sending an external signal to the molecular system. The molecules then behave like "molecular machines" or, at least, their prototypes. This field has experienced a spectacular development in the course of the last decade and books or special issues of chemistry journals have even been devoted to this new area of research [1–14].

The field of catenanes and rotaxanes [15–55] is particularly important in relation to molecular machines. Promising systems have been proposed which are based on redox reactions involving, in particular, formation or dissociation of organic acceptor-donor complexes [56] or of transition metal complexes [57–60]. Protonation/deprotonation, leading to dissociation and/or formation of given subcomplexes within a multicomponent structure (see [61] and references therein), is also an interesting possibility for inducing a motion within a multicomponent molecule. Among the many examples of molecular machines reported in the last decade, several examples of light-driven machines have been described [62–68]. Some of them contain a photoisomerizable group such as an azo benzene derivative. The light impulse converts the *trans* isomer to the *cis* isomer, leading to a significant change of the geometry of the photochemically active group, and thus strongly modifying its ability to interact with a given part of the molecular system. As a consequence, rearrangement may occur [69, 70]. Photoinduced electron transfer has also been used to set molecular systems in motion. Our group has been particularly interested in copper and, more recently, ruthenium(II)-containing interlocking or threaded ring systems [71–80]. Early work from our team, on the templated synthesis of catenanes using copper(I), has contributed in making these molecular systems reasonably accessible. It has also contributed in making catenanes, rotaxanes and, to a lesser extent, molecular knots, popular species that are nowadays considered to be normal and easy-to-make molecules. The reactions used to prepare the first copper(I)-complexed catenane (named "catenate") are represented in Figs. 1 and 2.

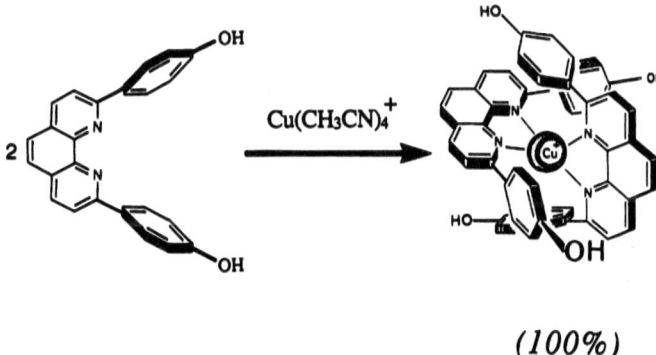

Fig. 1 A transition metal Cu(I) gathers and orients the two precursors of the [2]catenane

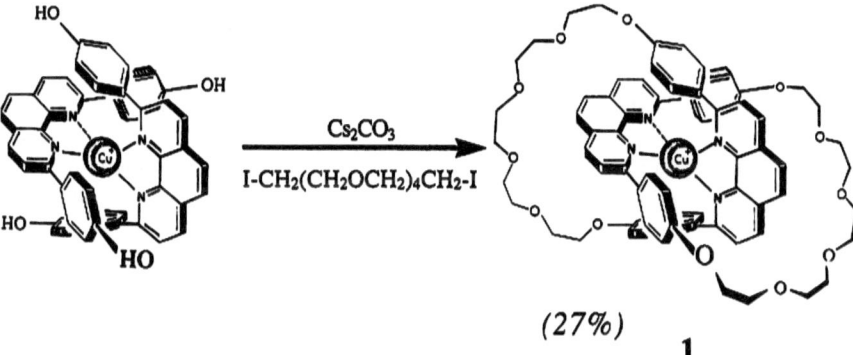

Fig. 2 Double cyclization reaction leading to the Cu(I) [2]catenane

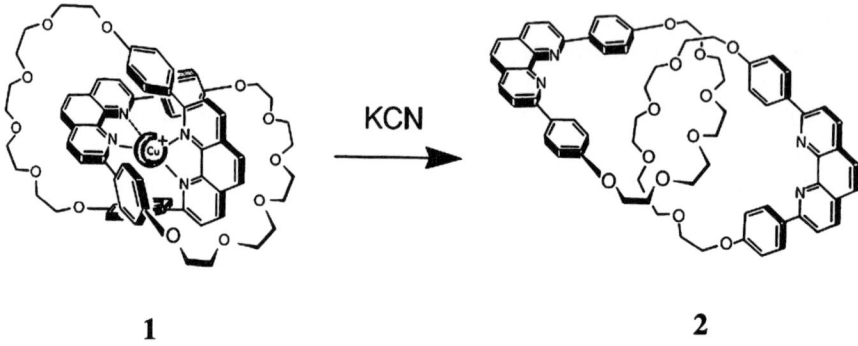

Fig. 3 Demetallation reaction to obtain the metal-free [2]catenane

By demetallation, a metal-free compound is obtained ("catenand"), as indicated in Fig. 3. The synthesis principle is very general and has been extended to numerous compounds, bearing various functionalities and containing between two and seven interlocking rings [81–83]. The synthesis of catenanes has also been extended to the preparation of porphyrinic rotaxanes and catenanes, displaying a very rich photochemistry in relation to intramolecular energy or electron transfer [84–86]. Recent work has also allowed the preparation of relatively simple interlocking-ring systems with yields close to 100% [87].

The many compounds, made and studied by our group, that behave as machine prototypes have been prepared via transition-metal-templated strategies, either using copper(I) or, more recently, ruthenium(II) as the central gathering core. These will be discussed in the following sections.

2
Electrochemically Driven Molecular Machines Based on the Copper(II/I) Couple

2.1
The First Catenane Whose Motions are Controlled by Oxidizing or Reducing the Central Copper Atom

The very first system made by our group for which a large amplitude motion can deliberately be triggered by an external signal was reported in 1994 [57]. Stoddart, Kaifer and coworkers reported their first molecular "shuttle" in the same year [56]. The compound made and studied in Strasbourg is a copper-complexed [2]catenane, displaying two very distinct coordination modes corresponding to two extremely different geometries. Having taken advantage of the roughly tetrahedral arrangement of the copper(I) center in bis-diimide complexes to construct catenanes and molecular knots [71–73, 88], the electrochemical properties of related copper complexes have later on been used in our group to set catenanes and rotaxanes in motion, playing on the two classical oxidation states of this metal, + 1 and + 2. The interconversion between both forms of the complex is electrochemically triggered and corresponds to the sliding motion of one ring within the other. It leads to a profound rearrangement of the compound and can thus be regarded as a complete metamorphosis of the molecule. The principle of the process is explained in Fig. 4.

The key feature of the transformation is the difference in preferred coordination number (*CN*) for the two different redox states of the metal: $CN = 4$ for copper(I) and $CN = 5$ (or 6) for copper(II). The organic backbone of the asymmetric catenate consists of a 2,9-diphenyl-1,10-phenanthroline (dpp) bidentate chelate included in one cycle and, interlocked to it, a ring containing two different subunits: a dpp moiety and a terdentate ligand, 2,2′ : 6′,2″-terpyridine (terpy). Depending upon the mutual arrangement of both interlocking rings, the central metal copper can be tetrahedrally complexed (two dpp) or 5-coordinate (dpp + terpy). Interconversion between these two complexing modes results from a complete pirouetting of the two-site ring. It can easily be induced electrochemically or by means of a chemical reductant or oxidant. From the stable tetrahedral monovalent complex, oxidation leads to a 4-coordinate Cu(II) state, which rearranges to the more stable

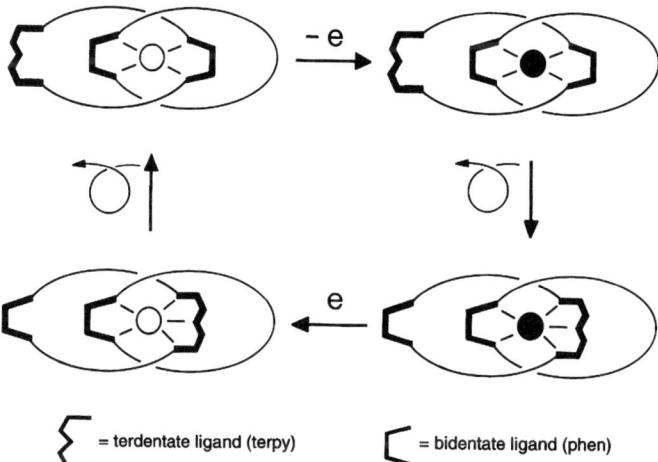

Fig. 4 Principle of electrochemically induced molecular motions in a copper complex catenane. The stable 4-coordinate monovalent complex (*top left*, the *white circle* represents Cu(I)) is oxidized to an intermediate tetrahedral divalent species (*top right*, the *black circle* represents Cu(II)). This compound undergoes a complete reorganization process to afford the stable 5-coordinate Cu(II) complex (*bottom right*). Upon reduction, the 5-coordinate monovalent state is formed as a transient (*bottom left*). Finally, the latter undergoes the conformational change that regenerates the starting complex

5-coordinate compound. The process can be reversed by reducing the divalent state to the 5-coordinate Cu(I) complex obtained as a transient species before a changeover process takes place to afford back the starting tetrahedral monovalent state. The real molecules are represented in Fig. 5 as well as the square scheme interconverting the 4- and the 5-coordinate species.

The synthetic strategy is derived from that represented in a schematic fashion in Figs. 1 and 2 for the archetype. By applying the principle of Fig. 4 to 3^+, it is clear that, for the same oxidation state (either Cu(I) or Cu(II)), the two forms should display significantly different physical properties. In particular, it is expected that the spectroscopic and electrochemical properties of the divalent copper catenates depend strongly on the coordination number of the metal. This could be demonstrated by oxidizing the copper(I) catenate 3^+ and subsequently monitoring the absorption spectrum and the redox properties of the divalent complex obtained as a function of time. 3_4^{2+}, the tetrahedral Cu(II) species (the subscript 4 indicates the coordination number) obtained immediately after oxidation either by Br_2 or via electrolysis, is a deep green complex in solution (λ_{max} = 670 nm, ε = 830 M^{-1} cm^{-1} in CH_3CN). This is in agreement with previous studies on the divalent form of the copper cate-

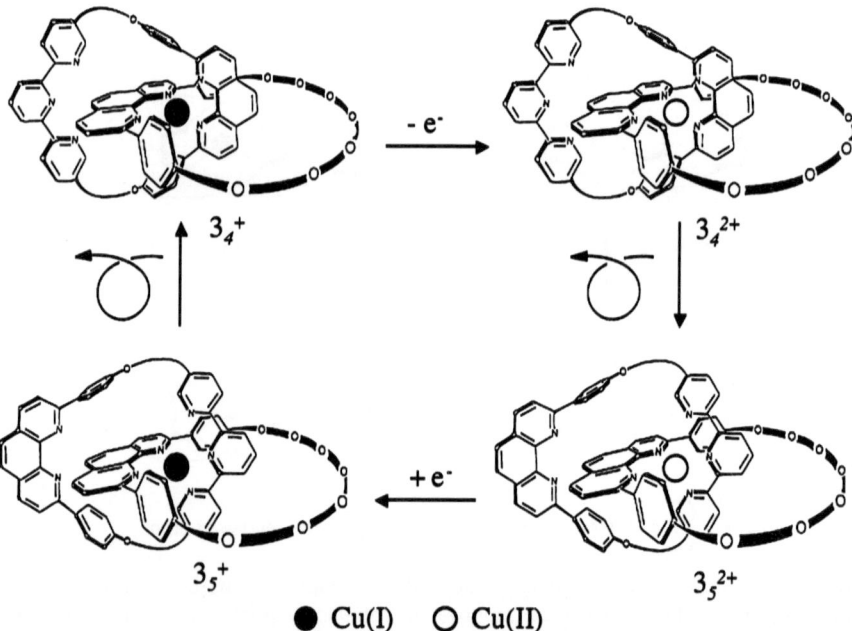

Fig. 5 Square scheme for catenane 3^{n+}. The subscripts 4 and 5 indicate the copper coordination number in each complex. The organic backbone is a [2]catenane, consisting of two different interlocking rings. One incorporates a single unit dpp, whereas the other cycle contains two coordinating fragments (a dpp unit and a terpy moiety)

nate of Fig. 3, ([Cu(II)2]$^{2+}$: λ_{max} = 685 nm, ε = 800 M^{-1} cm^{-1} in CH$_2$Cl$_2$). The electronic spectrum of the oxidized solution changes with time: a drastic intensity decrease around 670 nm is observed to afford a pale yellow-green complex. This slow process is in agreement with the changeover reaction represented in Fig. 3 and lead to the 5-coordinate copper(II) complex, 3_5^{2+} (from 3_4^{2+}), in which a coordinated dpp chelate has been replaced by the incoming terpy unit belonging to the same cycle. The spectral properties of the starting species and its isomeric product are in accordance with the coordination number and the geometry assumed around each copper center [89, 90].

Interestingly, the transformation of Fig. 5 is accompanied by a change in the electrochemical properties of the complex, paralleling the spectroscopic changes. As expected, the tetrahedral copper complex has a relatively high redox potential, $3_4^{2+/+}$: E_o = + 0.63 V vs. SCE in CH$_3$CN whereas the 5-coordinate species has a slightly negative potential, pointing to greater stabilization of the divalent copper than in the 4-coordinate species, $3_5^{2+/+}$: E_o = – 0.07 V.

Finally, reduction of 3_5^{2+} can quantitatively be carried out by electrolysis to regenerate the starting copper(I) complex, 3_4^+. Rearrangement of the monovalent complex ($3_5^+ \rightarrow 3_4^+$) takes place on the second time scale, as estimated from cyclic voltammogram (CV) measurements.

This first system corresponds more to the demonstration of a principle than to a practically promising molecular machine. The most obvious weak point of the complex is its kinetic inertness. It takes virtually hours, at room temperature in a coordinating solvent such as acetonitrile, for the thermodynamically unstable 4-coordinate complex 3_4^{2+} to rearrange and afford the stable form of the divalent copper complex, 3_5^{2+}. As already mentioned, the changeover process of the monovalent complex, ($3_5^+ \rightarrow 3_4^+$) is much faster than the reverse rearrangement on the divalent copper complex ($3_4^{2+} \rightarrow 3_5^{2+}$). The sliding process has to involve decoordination of the metal. This step is expected to be much slower for Cu(II) than for Cu(I) due to the greater charge of the former cation.

2.2
Copper-Complexed [2]Catenane in Motion with Three Distinct Geometries

Multistage systems seem to be uncommon, although they are particularly challenging and promising in relation to nanodevices aimed at important electronic functions and, in particular, information storage [91–94]. Among the few examples reported in recent years, three-stage catenanes are particularly significant since they lead to unidirectional rotary motors [10]. In the mid-1990s, our group described a particular copper-complexed [2]catenane that represents an example of such a multistage compound [95]. The molecule displays three distinct geometries, each stage corresponding to a different coordination number of the central complex (CN = 4, 5, or 6). The principle of the three-stage electrocontrollable catenane is represented in Fig. 6. Similarly to the simpler catenane 3^{n+} of Fig. 5, discussed above, it relies on the important differences in stereochemical requirements for coordination of Cu(I) and Cu(II). For the monovalent state the stability sequence is $CN = 4 > CN = 5 > CN = 6$. On the contrary, divalent copper is known to form stable hexacoordinate complexes, with pentacoordinate systems being less stable, and tetrahedral Cu(II) species being even more strongly disfavored.

The synthesis of the key catenate $[Cu(I)N_4]^+PF_6^- = 4_4^+$ (Fig. 7a) derives from the usual three-dimensional template strategy [71, 73].

The visible spectrum of this deep red complex shows a metal-to-ligand charge transfer (MLCT) absorption band ($\lambda_{max} = 439$ nm, $\varepsilon = 2570$ mol^{-1}L cm^{-1}, MeCN). Cyclic voltammetry of a MeCN solution shows a reversible redox process at + 0.63 V (vs. SCE). Both the CV data and the UV-vis spectrum are similar to those of other related species [57, 73]. The reaction of 4_4^+ with KCN afforded the free catenand (not represented), which was subsequently reacted with Cu(BF$_4$)$_2$ to give 4_6^{2+} as a very pale green complex. The hexacoordinate structure of this species was evidenced by UV-vis spectroscopy and electrochemistry. The CV shows an irreversible reduction at – 0.43 V (vs. SCE, MeCN). These data are similar to those obtained for the complex Cu(diMe-tpy)$_2$(BF$_4$)$_2$ (diMe-tpy = 5,5''-dimethyl-2,2' : 6',2''-terpyridine).

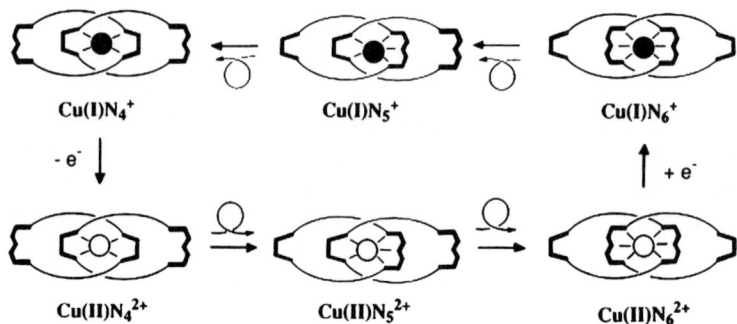

Fig. 6 A three-configuration Cu(I) catenate whose general molecular shape can be dramatically modified by oxidizing the central metal (Cu(I) to Cu(II)) or reducing it back to the monovalent state. Each ring of the [2]catenate incorporates two different coordinating units: the bidentate dpp unit is symbolized by a U whereas the terpy fragment is indicated by a stylized W. Starting from the tetracoordinate monovalent Cu complex Cu(I)N$_4^+$ (*top left*) and oxidizing it to the divalent state Cu(II)N$_4^{2+}$, a thermodynamically unstable species is obtained which should first rearrange to the pentacoordinate complex Cu(II)N$_5^{2+}$ by gliding of one ring (*left*) within the other and, finally, to the hexacoordinate stage Cu(II)N$_6^{2+}$ by rotation of the second cycle (*right*) within the first one. Cu(II)N$_6^{2+}$ is expected to be the thermodynamically stable divalent complex. The double ring-gliding motion following oxidation of Cu(I)N$_4^+$ can be inverted by reducing Cu(II)N$_6^{2+}$ to the monovalent state (Cu(I)N$_6^+$ (*top right*), as represented on the *top line* of the figure

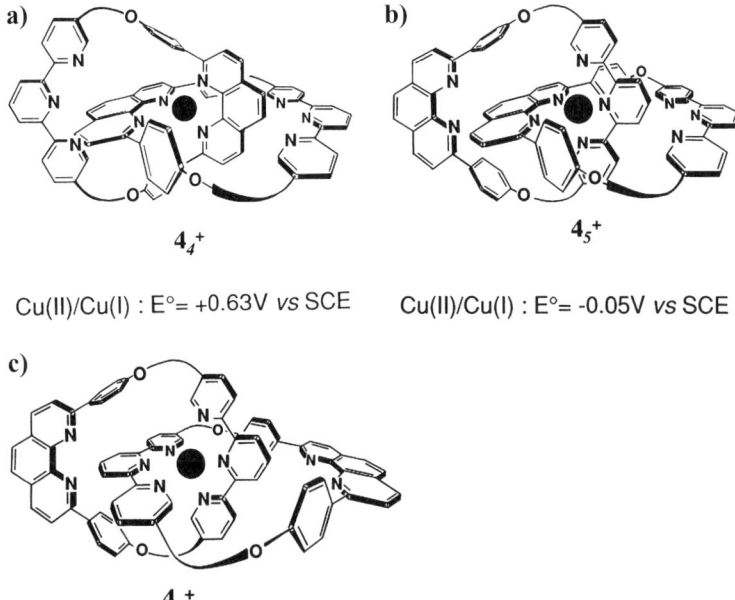

Fig. 7 Three forms of the copper-complexed catenane, each species being either a monovalent or a divalent complex: **a** 4-coordinate complex, **b** 5-coordinate complex, **c** 6-coordinate complex

When a MeCN dark red solution of 4_4^+ was oxidized by an excess of $NO^+BF_4^-$, a green solution of 4_4^{2+} was obtained. The CV is the same as for the starting complex, and the visible absorption spectrum shows a band at $\lambda_{max} = 670$ nm, $\varepsilon = 810$ mol^{-1} L cm^{-1}, in MeCN, typical of these tetrahedral Cu(II) complexes [57]. A decrease of the intensity of this band was observed when monitoring it as a function of time. This fact is due to the gliding motion of the rings to give the penta- and hexacoordinate Cu(II) complexes, whose extinction coefficients are lower than that of 4_4^{2+} (ca. 125 and 100, respectively). The rate of the rearrangement is roughly twice as large as that for the monoterpy-related catenate [57], which is in accordance with the presence of two terpy moieties. The final product is 4_6^{2+}, as indicated by the final spectro- and electrochemical data. A similar behavior was observed when a solution of 4_4^+ was electrochemically oxidized.

When either the 4_6^{2+} solution resulting from this process or a solution prepared from a sample of isolated solid $4_6^{2+}(BF_4^-)_2$ were electrochemically reduced at – 1 V, the tetracoordinate catenate was quantitatively obtained. The cycle depicted in Fig. 6 was thus completed. The changeover process for the monovalent species is faster than the rearrangement of the Cu(II) complexes, as previously observed for the above-mentioned related catenate 3^{n+} [57]. In fact, the rate is comparable to the CV time scale and three Cu species are detected when a CV of a MeCN solution of $4_6^{2+}(BF_4^-)_2$ is performed. The waves at + 0.63 V and − 0.41 V correspond, respectively, to the tetra- and hexacoordinate complexes mentioned above. By analogy with the value found for the previously reported copper-complexed catenane, leading to a 4- or 5-coordinated system, depending on the copper oxidation state (− 0.07 V for the 5-coordinate species) [57], the wave at − 0.05 V is assigned to the pentacoordinate couple (Fig. 7b). Both Cu(I) and Cu(II) pentacoordinate complexes are intermediates which could be characterized by various techniques as transient species, but in principle, the present system does not allow one to stop motions at this stage. Only three complexes of the six species of Fig. 6 have been spectroscopically identified.

The overall cyclic process depicted in Fig. 6 does not consume or produce electrons; it describes the interconversion between three molecular topographies (CN = 4, 5, or 6), the topological properties of the backbone being of course unmodified in the whole process.

2.3
A Molecular Shuttle: Gliding of a Ring on its Axle
Under the Action of an Electrochemical Signal

A [2]rotaxane is a molecular system consisting of a ring threaded by a string with two blocking groups attached at both ends of the string to prevent dethreading. Such compounds were first made some time ago [96, 97], but they have been mostly considered as chemical curiosities. Recently, rotaxanes underwent a real revival due to the newly developed efficient procedures that make them relatively easy to prepare [98–100], and also because of their electro- and photochemical properties [101–103] and their aptitude to undergo controlled molecular motions [56, 62, 104]. In previous work on catenanes, the gliding motion of a ring within the other was studied by setting in motion either one cycle or both rings [57, 63, 89, 95]. In the rotaxane discussed here, a ring is translated along a rod-like component on which it is threaded (Fig. 8). The motion is represented in a schematic fashion in Fig. 8.

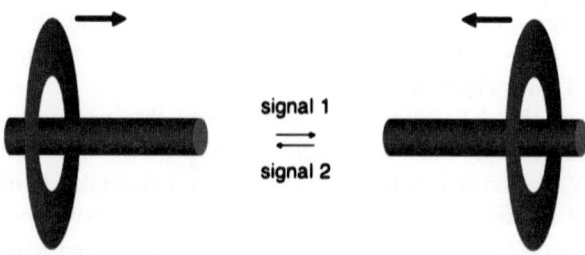

Fig. 8 Translation motion of a ring on an axle in a rotaxane

The main focus of the present review article is more to discuss the dynamic properties of the compounds than to describe their synthesis. We will thus only mention the synthetic principles leading to the various rotaxanes made by our group. The strategy relies on the ability of copper(I) to gather the two constitutive organic fragments (a ring incorporating a bidentate chelate and an open chain component) and to force the string to thread through the ring. This threading step is generally quantitative provided the stoichiometry of the reaction is carefully respected due to the selective formation of very stable tetrahedral copper(I) complexes. It can be extended to strings containing two different sites, such as bidentate and terdentate coordinating units. The 2-state copper-complexed rotaxane 5^{n+}, whose functioning is reminiscent of a molecular "shuttle" has been made following this principle.

This system functions on the same principle as that of the [2]catenanes 3^{n+} and 4^{n+} discussed in the preceding paragraph. Whereas a CN of four (usually with a roughly tetrahedral arrangement of the ligands) corresponds to stable monovalent systems, copper(II) requires higher coordination numbers. In the present case, the coordination number of five (square pyramidal or trigonal bipyramidal geometries) is the preferred situation of the copper(II) complex

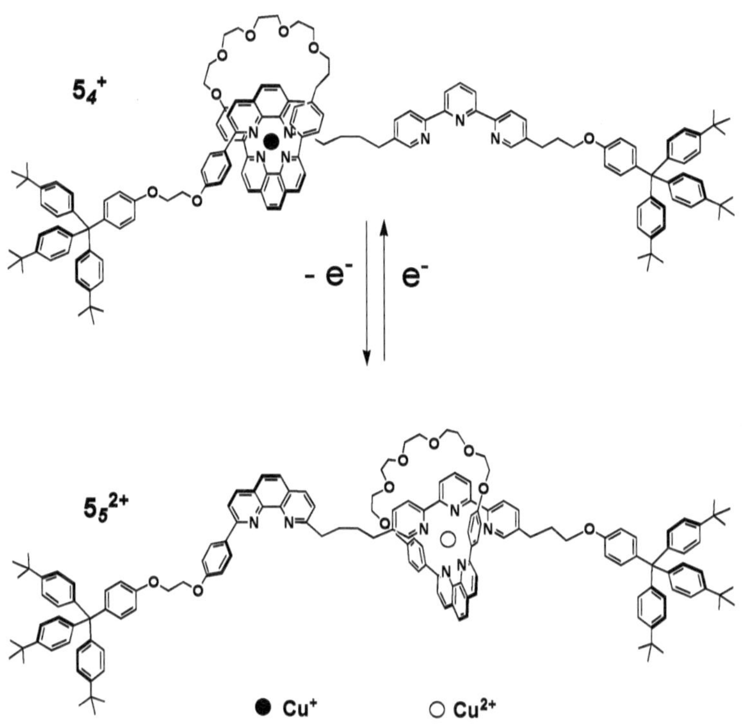

Fig. 9 Copper(I) rotaxane 5^{n+}: 4-coordinate (*upper line*) and 5-coordinate (*lower line*) situations

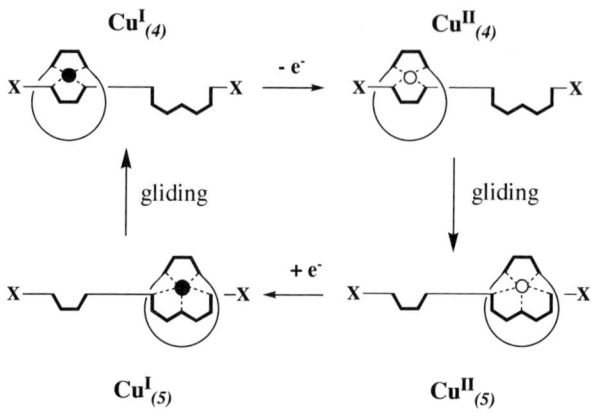

X = bulky stopper

Fig. 10 Square scheme in the threaded compounds. Subscripts 4 and 5 indicate the copper coordination number in each complex. The stable Cu(I)$_{(4)}$ complex is oxidized to an intermediate tetrahedral divalent species Cu(II)$_{(4)}$, which undergoes a rearrangement to afford the stable Cu(II)$_{(5)}$ complex. Upon reduction, a Cu(I)$_{(5)}$ species is formed as a transient, which finally reorganizes to regenerate the starting complex. The *black circle* represents Cu(I) and the *white circle* represents Cu(II)

5_5^{2+}. Changing the *CN* of the copper center from four to five, and vice versa, corresponds to gliding the ring from one position to the other. The square scheme illustrating this principle is given in Fig. 10.

The electrochemical behavior of 5^{n+} is particularly clean since only the 4- and the 5-coordinate geometries can be obtained by translating the metal-complexed ring from the phenanthroline site to the terpyridine site. These electrochemically induced molecular motions (Figs. 9 and 10) can be monitored by CV and controlled potential electrolysis experiments. From the CV measurements at different scan rates (from 0.005 to 2 V s^{-1}) both on the copper(I) and copper(II) species, it could be inferred that the chemical steps (motions of the ring from the phenanthroline to the terpyridine sites and vice versa) are slow on the time scale of the experiments. It is noteworthy that oxidation of 1+ to the divalent copper(II) state affords exclusively the 5-coordinate species after rearrangement of the system. The bis-terpyridine complex, which would be formed by decomplexation of the copper(II) center and recoordination to the terpyridine fragments of two different strings, is not detected. This observation is important in relation to the general mechanism of the changeover step converting a 4-coordinate Cu(II) species (5_4^{2+}) into the corresponding stable 5-coordinate complex (5_5^{2+}). It tends to indicate that the conversion does not involve full demetallation of Cu(II) followed by recomplexation. It is rather an intramolecular reaction, probably consisting of several elemental dissociation association steps involving the phenanthroline and terpyridine fragments of the string as well as solvent molecules and, possibly, counterions.

2.4
Transition-Metal-Containing Rotaxane in Motion: Electrochemically Induced Pirouetting of the Ring on a Threaded Dumbbell

In the previously discussed examples, gliding motion of a ring around another one in the case of a catenanes or translation motion along a molecular thread in the case of rotaxane 5^{n+} were studied. The present system [59] is based on a rotaxane in which another type of motion, namely the pirouetting of the wheel around its axle, can be electrochemically triggered, as represented in a very schematic fashion in Fig. 11.

The driving force of this motion is here again based on different geometrical preferences for Cu(I) and Cu(II). The wheel of the rotaxane is a bis-coordinating macrocycle containing both a bidentate moiety, a phen nucleus, and a terdentate unit, a terpy fragment. The axle incorporates only one bidentate moiety, a 2,9-diphenyl-1,10-phenanthroline (dpp) motif.

The strategy used to prepare the rotaxane is based on the utilization of a monostoppered species, which will be first threaded through the ring to form the pre-rotaxane intermediate and, second, to add the second stopper afterwards (Fig. 12).

The main advantage of this method is to limit dethreading of the macrocycle during the stoppering reaction. The rotaxane 6_4^+ was synthesized following this strategy. The two forms of the rotaxane, 6_4^+ and 6_5^{2+} (4- or 5-coordinate complex) are represented on Fig. 13.

As expected, an intense MLCT absorption band was observed for 6_4^+, in the visible range (λ_{max} = 438 nm, ε = 2830 M^{-1} cm^{-1}), by analogy with Cu(dpp)$_2^+$-based complexes.

The electrochemical behavior of 6_4^+ (CH$_2$Cl$_2$/CH$_3$CN solution) has been studied by CV. As for the corresponding catenanes, it was clearly demon-

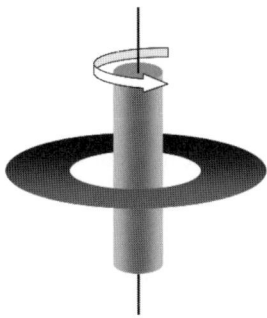

Fig. 11 Rotation of the axle within the ring in a rotaxane

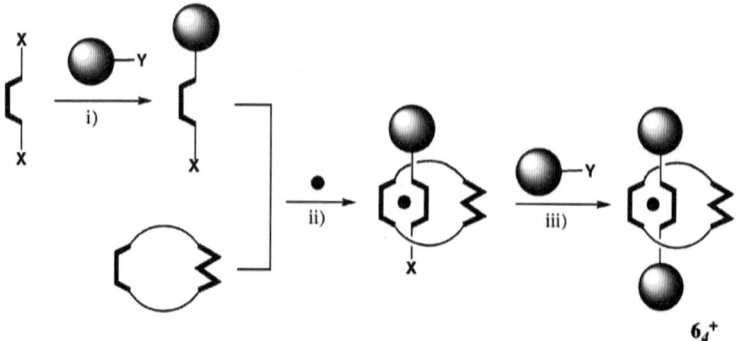

Fig. 12 Principle of synthesis of rotaxane 6_4^+ (see Fig. 13). *i* Formation of a monostoppered axle, and *ii* threading of the axle through the wheel. *iii* Dethreading of the macrocycle is prevented by adding a second blocking group

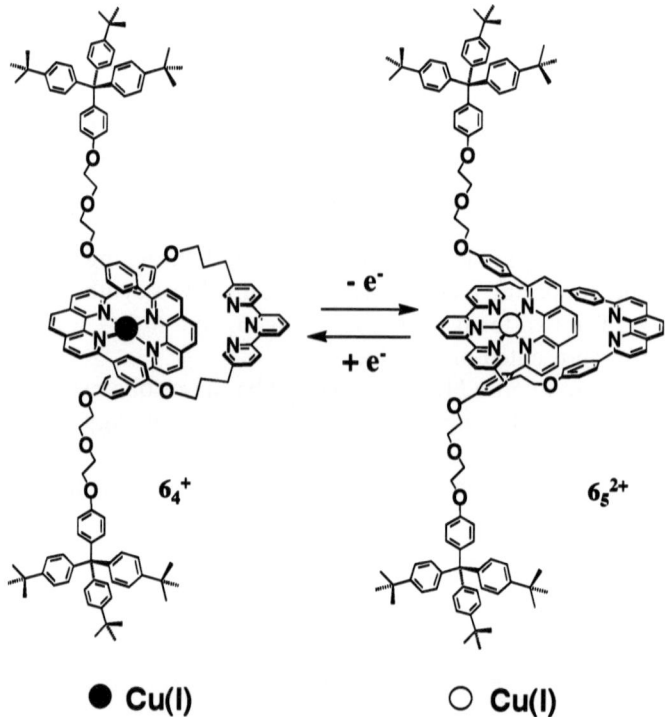

Fig. 13 Representation of the two stable forms of the rotaxane 6^{n+}

strated that the compound undergoes a complete rearrangement upon reduction or oxidation of the copper center, going from a 4-coordinate system for copper(I) to a pentacoordination mode for divalent copper. The rate of the rearrangement process is particularly important to know, and this is why a kinetic study was undertaken, based on relatively old electrochemical methods developed by Shain and Nicholson [105]. Without entering into the details of the technique, it is based on CV measurements performed at various scan rates. It was thus possible to measure with a relatively high accuracy the rate constants for the rearrangement reaction of the thermodynamically unstable compounds, namely the 4-coordinate copper(II) complex 6_4^{2+} and the 5-coordinate copper(I) species 6_5^+. The half-life time of the pentacoordinate Cu(I) rotaxane is about 56 ms and that of the 4-coordinate divalent copper complex 6_4^{2+} is in the range of 2 min. The half-life times found for these complexes are thus significantly shorter that those of the corresponding complexes in the catenane series (compare 56 ms and a few seconds for the lifetime of the pentacoordinated Cu(I) complex 3_5^+ or minutes to hours for the tetracoordinated copper(II) species 3_4^{2+}). The "pirouetting" motion is represented in a schematic manner in Fig. 13. The overall electrochemically induced rearrangement is summarized in Fig. 14, with the rate constants of the corresponding motions.

Once again, the detailed electrochemical study of the present two-geometry rotaxane (6_4^+ or 6_5^{2+}) underlines the noticeable difference in the

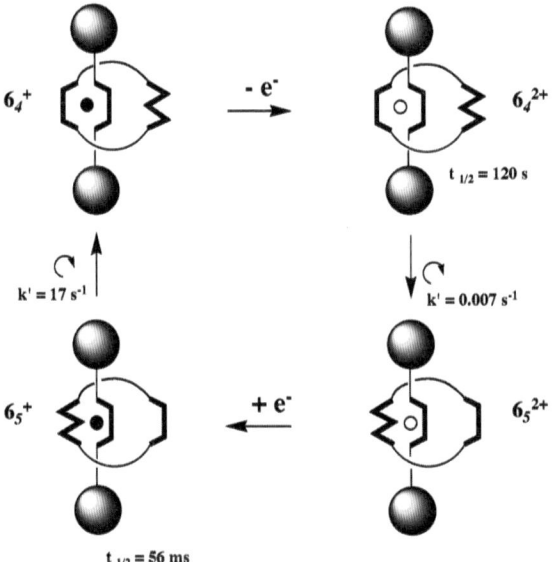

Fig. 14 Principle of the electrochemically induced molecular motions in a copper complex rotaxane. The stable 4-coordinate monovalent complex is oxidized to an intermediate tetrahedral divalent species. This compound undergoes a rearrangement to afford the stable 5-coordinate copper(II) complex. Upon reduction, the 5-coordinate monovalent state is formed as transient. Finally, the latter undergoes the reorganization process that regenerates the starting complex (the *black circle* represents Cu(I) and the *white circle* represents Cu(II))

kinetic rate constants for the reorganization processes of the transitory 4-coordinate copper(II) form (slow) and 5-coordinate copper(I) complex (fast). Divalent copper complexes are much slower to rearrange than the monovalent ones. In addition, and this is certainly the main conclusion of the electrochemical study on the presently discussed rotaxane, the pirouetting movement of the macrocycle around its axle is much faster than the translation of a macrocycle along a molecular thread or the swinging motion of a ring in a two-shape catenane. It is noteworthy that for the latter case, the coordinating fragments around the metal are strictly identical, the only difference lying in the rotaxane nature of the compound compared to the catenane topology of the previously elaborated molecule.

2.5
Fast-Moving Electrochemically Driven Machine Based on a Pirouetting Copper-Complexed Rotaxane

The rate of the motion in artificial molecular machines and motors is obviously an important factor. Depending on the nature of the movement, it can range from microseconds, as in the case of organic rotaxanes acting as light-driven molecular shuttles [65], to seconds, minutes or even hours in other systems involving threading–unthreading reactions [67, 68] or metal-centered redox processes based on the Cu(II)/Cu(I) couple (as in the case of 3^{n+}, for instance) [57].

In order to increase the rate of the motions, a new rotaxane in which the metal center is as accessible as possible was prepared, the ligand set around the copper center being thus sterically hindering compared to previously related systems. Ligand exchange within the coordination sphere of the metal is thus facilitated as much as possible. The two forms of the new bistable rotaxane, 7_4^+ and 7_5^{2+}, are depicted in Fig. 15 (as usual, the subscripts 4 and 5 indicate the coordination number of the copper center). The molecular axis contains a "thin" 2,2'-bipyridine motif, which is less bulky than a 1,10-phenanthroline fragment and thus is expected to spin more readily within the cavity of the ring. In addition, the bipy chelate does not bear substituents in α position to the nitrogen atoms. 7_4^+ rearranges to the 5-coordinate species 7_5^{2+} after oxidation and vice versa. The electrochemically driven motions were studied by CV. The CV data are shown in Fig. 16, for two different scan rates (CV in MeCN with 0.1 mol L^{-1} Bu$_4$NBF$_4$). At a relatively low scan rate (100 mV/s), both rearrangement processes are faster than the potential sweep. By contrast, if the sweep rate is increased up to several volts per second, the return wave for the 4-coordinate copper complex is observed. By varying the scan rate and by analyzing the CV curves obtained, it is even possible to determine the rate of the rearrangement process for both the 4-coordinate complex and the 5-coordinate form. The curve *a* represents 20 cycles, showing that the chemical reversibility of the process is also high.

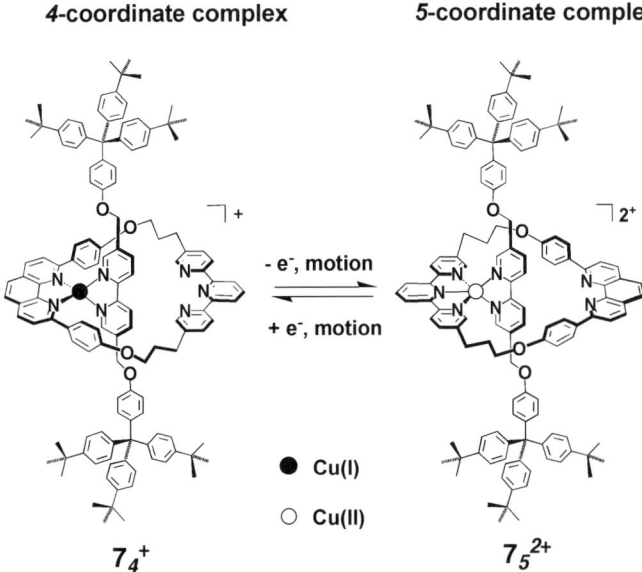

Fig. 15 Electrochemically induced pirouetting of the ring in rotaxane 7^{n+}. The bidentate chelate and the tridentate fragment are alternatively coordinated to the copper center

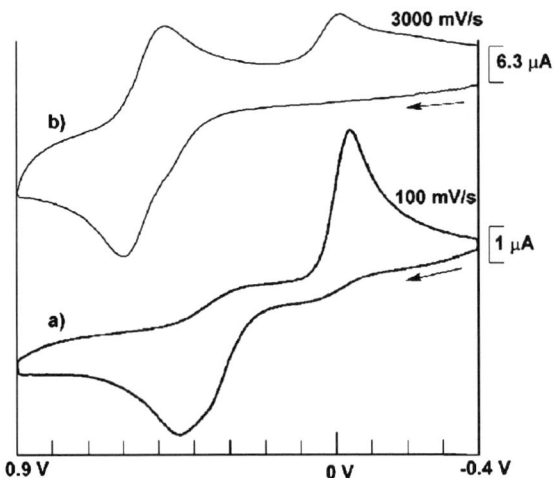

Fig. 16 CVs of 7^{n+} in MeCN with 0.1 mol L^{-1} Bu$_4$NBF$_4$ at two different scan rates. *a* Represents 20 cycles and shows that the various CV curves are superimposable. The potentials are referenced versus a silver quasi-reference electrode. *b* Represents a CV at a sweep rate of 3000 mV/s and shows the return wave for the 4-coordinate copper complex

A lower limit for the rate constant k of the process can be estimated as $> 500 \text{ s}^{-1}$ (or $\tau < 2$ ms, with $\tau = k^{-1}$).

$$7_5^+ \xrightarrow{k > 500 \text{ s}^{-1}} 7_4^+ \tag{1}$$

The rearrangement rate for the 4-coordinate Cu(II) complex is smaller than for the monovalent complex. It is nevertheless several orders of magnitude larger than in related catenanes or rotaxanes with more encumbering ligands:

$$7_4^{2+} \xrightarrow{5 \text{ s}^{-1}} 7_5^{2+} \tag{2}$$

This last example shows that subtle structural factors can have a very significant influence on the general behavior (rate of the movement, in particular) of copper(II/I)-based molecular machines. Further modifications will certainly lead to new systems with even shorter response times.

3
Molecular Motions Triggered by an External Chemical Signal

3.1
Complete Rearrangement of a Multi-porphyrinic Rotaxane by Metallation–Demetallation of the Central Coordination Site

In the present system, we will discuss the dynamic properties of a multiporphyrinic [2]rotaxane in which a gold(III) porphyrin is part of the ring and two free-base or zinc(II) porphyrins play the role of the stoppers. Rotation of the string-like fragment within the ring between two diametrically opposed positions is triggered by metallation–demetallation of the central coordination site. This dynamic process will bring to close proximity, or spread a long distance apart, given porphyrinic components of the system. The principle is depicted in Fig. 17.

The compound made and studied is a [2]rotaxane in which the string-like fragment bears two zinc(II) porphyrins as blocking groups. The ring through which the string is threaded incorporates a gold(III) porphyrin. These metalloporphyrins are key components of multichromophoric systems undergoing photo induced electron transfer and proposed as models of given fragments of the photosynthetic reaction center [84, 101, 106].

We will not discuss the synthetic aspects of the present system [107, 108]. Rotaxane 8^+ was synthesized following a strategy previously developed in our group for making porphyrin-stoppered rotaxanes [84, 101].

The conformation of the Cu(I) complex 8^+ is indeed similar to that suggested in Fig. 18. In particular, NOE effects measured on $H_{5,6}$ and H_{py} demonstrate unambiguously that a close proximity exists between the rear of the 1,10-phenanthroline nucleus belonging to the dumbbell-like fragment

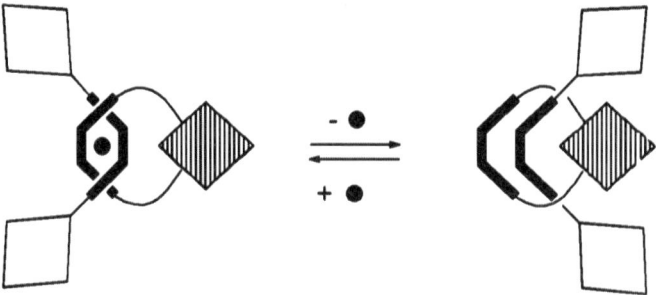

Fig. 17 A "pirouetting" machine in action: the mutual arrangement between the gold porphyrin (PAu$^+$ incorporated in the ring, *black diamond*) and the zinc porphyrins (PZn end function of the dumbbell, *white diamond*) is controlled by complexation/decomplexation of a metal center (*black circle*) within/from the central coordination site. In the complex, the gold porphyrin is remote from the two zinc porphyrins. After removal of the central metal, weak forces favor an attractive interaction between PAu$^+$ and the PZn nuclei, leading to a situation in which PAu$^+$ is pinched between the two PZn units. The interconversion between both situations implies a half-turn rotation of the threaded fragment ("axle") within the ring ("wheel")

and the ring-embedded porphyrin. As indicated in Fig. 18, demetallation of **8$^+$** affords **9**, this compound displaying a profoundly modified geometry as compared to **8$^+$**. In particular, NOE effects show close proximity between H_m and $H_{o'}$ as well as between H_{py} and $H_{o'}$ and between $H_{o'}$ and H_{Me}, which indicates that the geometry of the molecule is roughly as depicted in Fig. 18. Space-filling models suggest that within the demetallated rotaxane **9**, free rotation of the "axle" within the ring can take place. The driving force for bringing PAu$^+$ between the PZn units, playing the role of two jaws, is certainly related to the extremely different and complementary electronic properties of PAu$^+$ (electron acceptor) and PZn (electron donor). Very approximate geometrical features can be estimated from the models. Of particular interest are the center-to-center (Au$\cdots$Zn) and the edge-to-edge distances between PAu$^+$ and PZn. The estimated center-to-center separation is ca. 19 and ca. 7 Å for **8$^+$** and **9**, respectively. The edge-to-edge distance, which is more relevant to electron transfer, is ca. 12 and ca. 5 Å for **8$^+$** and **9**, although it should be kept in mind that **9** is certainly very flexible. It is thus difficult to estimate interatomic distances with high accuracy.

Interestingly, the interconversion between **8$^+$** and **9**, although leading to dramatic geometrical changes, is quantitative and reversible. The present system represents some kind of "pirouetting" machine but, obviously, it can by no means be regarded as a rotary motor since directionality is not controlled.

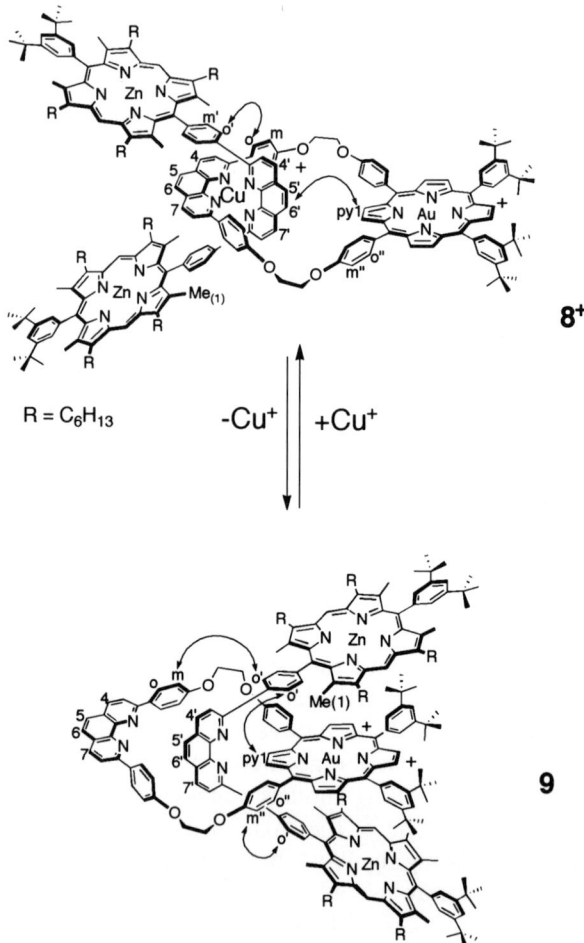

Fig. 18 Metallation–demetallation of the rotaxane induces a complete changeover of the molecule. The most important proton connectivities, as determined by 2D ^{1}H NMR, are indicated by *double arrows*

3.2
Use of a Chemical Reaction to Induce the Contraction/ Stretching Process of a Muscle-Like Rotaxane Dimer

Linear machines and motors are essential in many biological processes such as, in particular, contraction and stretching of the skeletal muscles. In relation to "artificial muscles", one-dimensional molecular assemblies able to undergo stretching and contraction motions thus represent an exciting target.

A multicomponent system able to contract or stretch under the action of an external chemical signal was designed and made by our group a few years ago. The system is based on a symmetrical doubly threaded topology as represented in Fig. 19. The motion is easy to visualize: both "strings" (mimicking the myosin-containing thick filament and the actin thin filament of the striated muscle) move along one another but stay together thanks to the rotaxane nature of the system.

The copper-complexed rotaxane dimer 10^{2+} was synthesized (more than 20 steps from commercially available compounds). As shown in Fig. 20, each "filament" contains both a bidentate chelate (coordinated to Cu(I) in compound 10^{2+}) and a tridentate chelate of the terpy type, which is free in the Cu(I) complex 10^{2+}. The rotaxane dimer was set in motion by exchanging the complexed metal centers. The free ligand, obtained in quantitative yield by reacting the 4-coordinate Cu(I) complex 10^{2+} (stretched geometry) with an excess of KCN, was subsequently remetallated with $Zn(NO_3)_2$ affording quantitatively the 5-coordinate Zn^{2+} complex 11^{4+} in the contracted situation (Fig. 20). The reverse motion, leading back to the extended situation 10^{2+}, could be easily induced upon addition of excess $Cu(CH_3CN)_4^+$. From CPK model estimations, the length of the organic backbone changes from 85 to 65 Å between both situations [109, 110].

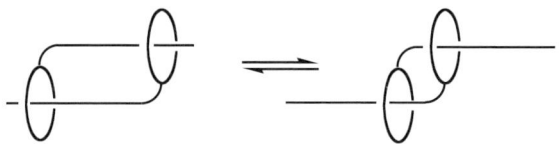

Fig. 19 Gliding of the filaments in a rotaxane dimer: interconversion of the stretched geometry and the contracted conformation

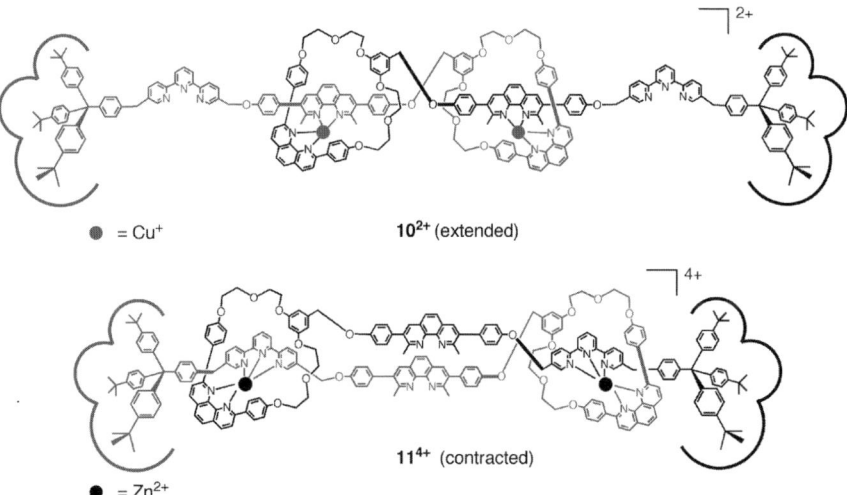

Fig. 20 The two states of the muscle-like molecule

4
Ruthenium(II)-Complexed Catenanes and Rotaxanes: Light-Driven Molecular Machines Based on Dissociative Excited States

4.1
Use of Dissociative Excited States to Set Ru(II)-Complexed Molecular Machines in Motion: Principle

Our group has recently described multicomponent ruthenium(II) complexes in which one part of the molecule can be set in motion photochemically [111, 112]. Among the light-driven molecular machine prototypes which have been described during the last few years, a very distinct family of dynamic molecular systems takes advantage of the dissociative character of ligand-field states in Ru(diimine)$_3^{2+}$ complexes [113–118]. In these compounds, one part of the system is set in motion by photochemically expelling a given chelate, the reverse motion being performed simply by heating the product of the photochemical reaction so as to regenerate the original state. In these systems, the light-driven motions are based on the formation of dissociative excited states. Complexes of the [Ru(diimine)$_3$]$^{2+}$ family are particularly well adapted to this approach. If distortion of the coordination octahedron is sufficient to significantly decrease the ligand field, which can be realized by using one or several sterically hindering ligands, the strongly dissociative ligand-field state (^{3}d-d state) can be efficiently populated from the metal-to-ligand charge transfer (^{3}MLCT) state to result in expulsion of a given ligand. The principle of the whole process is represented in Fig. 21.

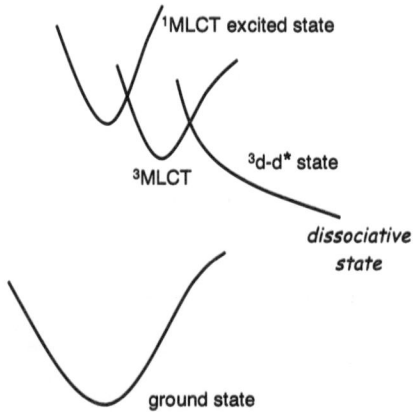

Fig. 21 The ligand-field state ^{3}d-d* can be populated from the ^{3}MLCT state, provided the energy difference between these two states is not too large. Formation of this dissociative state leads to dissociation of a ligand

It is thus essential that the ruthenium(II) complexes, which are to be used as building blocks of the future machines, contain sterically hindering chelates so as to force the coordination sphere of the metal to be distorted from the perfect octahedral geometry. We will discuss the synthesis of a rotaxane and a catenane of this family and briefly describe the photochemical reactivity of these molecules. The complexes made and studied incorporate encumbering ligands, which indeed facilitated the light-induced motions.

4.2
Templated Synthesis of a Pseudo-Rotaxane with a [Ru(diimine)$_3$]$^{2+}$ Core and its Light-Driven Unthreading Reaction

A rotaxane containing a ruthenium bis-phen complex (phen: 1,10-phenanthroline), acting as an axis, and a macrocycle incorporating a 2,2′-bipyridine (bipy) unit, threaded by the axis, has been synthesized. It was recently reported that a [Ru(diimine)$_2$]$^{2+}$ moiety can be inscribed in an axial compound by appropriate substitution of the diimine [119, 120]. Subsequently, the [Ru(diimine)$_2$]$^{2+}$-containing axial fragment has also been incorporated in a pseudo-rotaxane and even in a complete rotaxane, the threaded ring also being coordinated to the ruthenium(II) center through a diimine chelate unit [77, 121].

The synthetic strategy consists of a threading step (the axial component is threaded through the ring) followed by a stoppering reaction (a bulky substituent is attached at each end of the axis). The threading reaction was first tested on a model whose axis is end-functionalized by two unreactive chemical groups (ether functionalities; see Fig. 22). Complex 12-(PF$_6$)$_2$ is a yellow solid that is formed quantitatively from its dichloro precursor (purple complex) by replacing the Cl$^-$ ligands by CH$_3$CN in H$_2$O/CH$_3$CN. The macrocyclic compound 13, which incorporates a 2,2′-bipyridine ligand substituted at its 6- and 6′-positions by alkyl groups, and a dimethyldi(p-alkoxyphenyl)methane fragment derived from "bisphenol A", has been obtained by reaction of the suitable dibromo precursor (6,6′-di[2-(2-bromoethoxy)ethoxypropyl]- 2,2′-bipyridine) with the bisphenol A in 45% yield. Compound 13 has a 35-membered ring, and CPK models indicate that its size should be sufficient to allow the threading reaction outlined in Fig. 22, although the rotaxane-like molecule obtained should be tight, with contacts between the bisphenol A motif of the ring and the $-CH_2-CH_2-C_6H_4-CH_2-CH_2-$ fragment of the axial component. This steric hindrance may in part explain the poor yield of the reaction: 12^{2+} and 13 react in ethylene glycol (140 °C, 4 h, stoichiometric, concentration of the reactants 0.01 mol L^{-1}) to afford a 20–25% yield of a mixture of complexes containing 14^{2+} and 14$^{′2+}$ (orange solid) after chromatography (Fig. 22). This mixture displays only one round-shaped spot on a thin-layer chromatography, making the separation of these two complexes extremely difficult. By

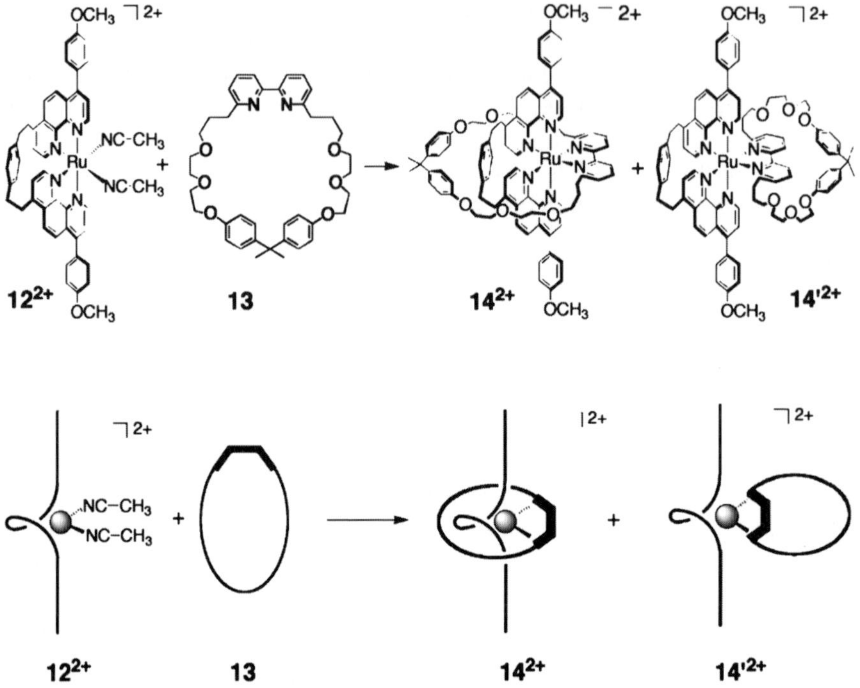

Fig. 22 Synthesis of the pseudo-rotaxane 14^{2+} and its *exo*-isomer $14'^{2+}$

comparison, 12^{2+} reacts with the acyclic ligand 6,6'-dimethyl-2,2'-bipyridine (6,6'-dmbp) to afford the corresponding complex in quantitative yield, under reaction conditions similar to those used for preparing 14^{2+}.

Single crystals of 14^{2+}-$(PF_6)_2$ could be obtained by slow diffusion of hexane in a solution of the complex in acetone, and an X-ray structure was obtained. As shown in Fig. 23, 14^{2+} is indeed a threaded species with a helical axis, the bis-1,10-phenanthroline ligand being wrapped around the metal center in a way similar to that recently observed with non-rotaxane-like species [119, 120]. The metal center is octahedrally coordinated, with little distortion. The Ru-N distances and N-Ru-N angles have the expected values (Ru-N distance was 2.055–2.068 Å for the phen ligands and 2.12–2.13 Å for the bipy part). The most striking feature of the structure is the distortion of the ring from planarity. Clearly, the ring is too small to accommodate "comfortably" the relatively thick axle and it can not run around the $-CH_2-CH_2-C_6H_4-CH_2-CH_2-$ part of the helical axis. The folded conformation of the macrocyclic component of 14^{2+} results in a non-symmetrical situation for which the "upper" and the "lower" parts of the rotaxane become non-equivalent in the solid state and in solution at low temperature, as evidenced by the ^{1}H NMR study. The very congested situation in 14^{2+}, as evidenced by the X-ray structure, tends to explain why the preparative yield is poor. The presence of a certain propor-

Fig. 23 X-ray structure of the Ru(II)-complexed pseudo-rotaxane

tion of the non-threaded species (Fig. 22) is also understood: the "unnatural" conformation of the ring in this species may be unfavorable but this destabilization energy is compensated by that introduced by the steric repulsion between the ring and the thread in the pseudo-rotaxane 14^{2+}. As expected, visible light irradiation of a solution of the $14^{2+}+14'^{2+}$ mixture in acetonitrile leads quantitatively to the dethreading products 12^{2+} and 13 ($\lambda > 400$ nm). The photochemical reaction can easily be monitored by UV-vis spectroscopy. The mixture of isomeric complexes has an absorption spectrum characteristic of $[\text{Ru}(\text{diimine})_3]^{2+}$ complexes, with a metal-to-ligand charge-transfer (MLCT) absorption band centered at 461 nm. Under irradiation, this band is gradu-

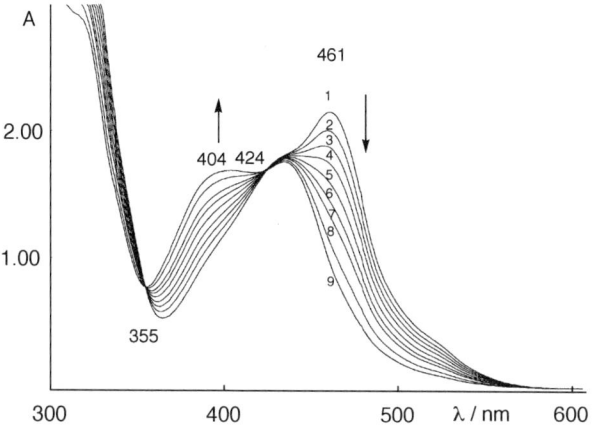

Fig. 24 Electronic absorption spectra in CH_3CN of the mixture of 14^{2+} and $14'^{2+}$ before (1) and after different irradiation times 2 20, 3 40, 4 60, 5 90, 6 120, 7 150, 8 210, and 9 300 s

ally replaced by the MLCT band of 12^{2+} (λ_{max} = 404 nm). Isosbestic points are observed at 355 nm and 424 nm as shown in Fig. 24.

The synthesis of the full rotaxane, containing two large stoppers, was subsequently carried out. Nevertheless, the poor yield of threaded compound and our failure to separate the threaded complex from its non-threaded isomer precluded any further study. Recent work has been performed which shows that by controlling in a better way the geometry of the ring, in particular by synthesizing a rigid analog of **13**, it is possible to obtain an endo-complex (threaded structure) exclusively [121]. These new complexes are obviously much more promising as light-driven molecular machines than compounds derived from the small and flexible ring **13**.

4.3
Construction of a [2]Catenane Around a Ru(diimine)$_3^{2+}$ Complex Used as Template

The design of the system and the synthetic strategy are depicted in Fig. 25. The main point of the design is the observation that it should be possible to incorporate two bidentate chelates of the octahedron in a ring and subse-

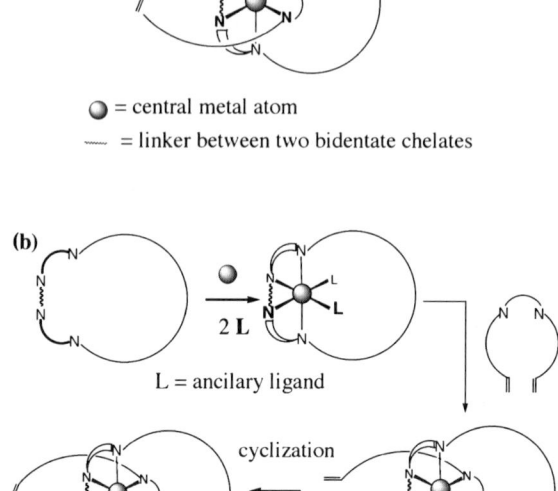

Fig. 25 a Schematic representation of a transition-metal-complexed [2]catenane containing two different rings. One of the macrocycles incorporates a bidentate chelate, whereas the other contains two bidentate coordinating fragments with a *cis* arrangement. **b** Synthetic strategy

quently to thread a fragment containing the third chelate through the ring. This second process would of course be driven by coordination to the central metal.

Tetradentate ligands consisting of two separate bidentate ligands connected by an appropriate spacer and leading to C_2-symmetric complexes have already been reported. A particularly interesting example is that of von Zelewsky's chiragens [122], consisting of two chiral bipy derivatives. Our group has also proposed a bis-phen molecule leading to a Ru(phen)$_3^{2+}$-derivative with a clearly identified axis bearing chemical functions [119, 120]. The substitution positions on the phen nuclei attached to the functions to be used for further derivatization are different than those corresponding to the previous axis-containing complex, as shown in Fig. 26. They seem to be appropriate to the formation of cyclic complexes.

The synthetic procedure starts with the preparation of a large ring incorporating two phen units. The choice of ring was dictated by CPK models and

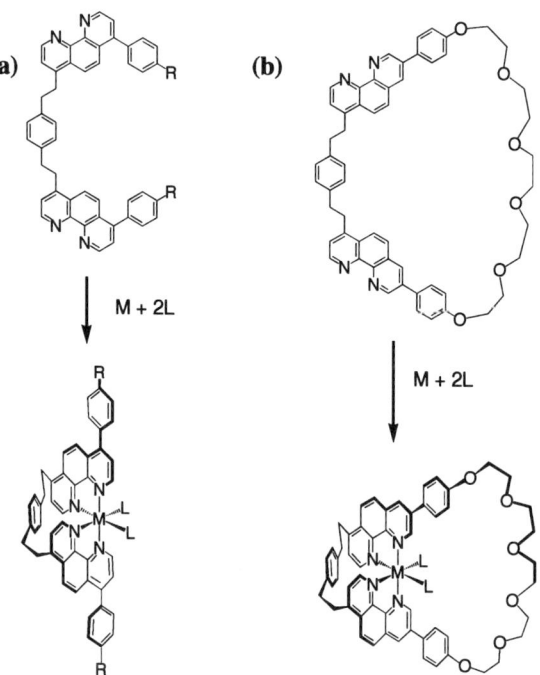

Fig. 26 Formation of an axial (**a**) or a macrocyclic (**b**) complex. In both cases, connection of two positions *para* to the N atoms of the phen nuclei by the $-CH_2CH_2-C_6H_4-CH_2CH_2-$ bridge leads to a *cis* arrangement. Introduction of aromatic groups $(-C_6H_4-R)$ on the other *para* positions leads to the axial complex (**a**), whereas the macrocyclic complex (**b**) can be obtained by utilizing the *meta* positions (C8) to attach the $-C_6H_4-R$ aromatic groups

by synthesis considerations. The precursors and the open-chain and cyclic compounds incorporating two phen fragments are represented in Fig. 27.

15 was prepared in four steps from 3-bromo-7-methyl-1,10-phenanthroline. It is a 50-membered ring which, in CPK models, looks adapted to the formation of octahedral bis-phen complexes, the two phen fragments being disposed *cis* to one another in the metal coordination sphere. Interestingly, the substitution positions of the *p*-alkoxyphenyl groups (8 and 8' in **15**) are determining. By contrast, if *p*-anisyl groups are introduced para to the N-atoms of the phen nuclei (positions 7 and 7'), wrapping the corresponding ligand around an octahedron leads to a system with a clearly identified axis [119, 120]. A key step is the coordination reaction, supposed to lead to the cyclic complex (Fig. 27). Several first-row transition metals were tested, leading to limited success. However, ruthenium(II) afforded the desired complex.

Complex 16^{2+} was formed by reacting **15** and $Ru(DMSO)_4Cl_2$, followed by refluxing the dichloro intermediate complex in CH_3CN-H_2O. $16^{2+}-(PF_6)_2$

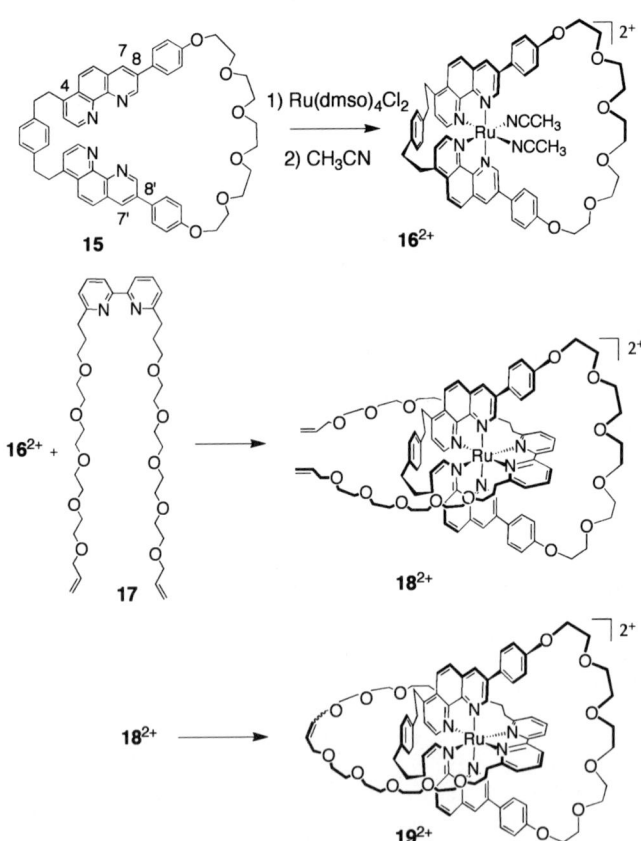

Fig. 27 Sequence of reactions affording the ruthenium (II)-complexed [2]catenane 19^{2+}

was obtained as an orange solid in 21% yield, after anion exchange. 16^{2+} is a rare example of a bis-phen or, more generally, a bis-bidentate octahedral complex with a *cis* arrangement, inscribed in a ring. The next step was carried out using 17 and the macrocyclic complex 16^{2+}. Threading of the "filament" 17 does take place under relatively harsh conditions (ethylene glycol: 140 °C) and the catenane precursor 18^{2+} was obtained in good yield (56%). The final compound, catenane 19^{2+}, was prepared from 18^{2+} in 68% yield by ring closing metathesis (RCM). The synthetic procedure used and the yield obtained were similar to those corresponding to the preparation of other transition-metal-containing catenanes and knots using a related RCM-based approach [88, 123]. 19^{2+}-$(PF_6)_2$ is a red-orange solid which has been fully characterized by various spectroscopic techniques. The ES-MS and ^{1}H NMR spectra afford clear evidence for the structure of 19^{2+}.

4.4
Photoinduced Decoordination and Thermal Recoordination of a Ring in a Ruthenium(II)-Containing [2]Catenane

The [2]catenane 19^{2+} was synthesized as described in the previous paragraph. The other [2]catenane of Fig. 28, 20^{2+}, was prepared using a slightly different procedure. Compound 19^{2+} consists of a 50-membered ring which incorporates two phen units and a 42-membered ring which contains the bipy chelate. Compound 20^{2+} contains the same bipy-incorporating ring as 19^{2+}, but the other ring is a 63-membered ring. Clearly, from CPK model considerations, 20^{2+} is more adapted than 19^{2+} to molecular motions in which both constitutive rings would move with respect to one another since the situation is relatively tight for the latter catenane. The light-induced motion and the thermal back reaction carried out with 19^{2+} or 20^{2+} are represented in Fig. 28. They are both quantitative, as shown by UV-vis measurements and by ^{1}H NMR spectroscopy.

The photoproducts, [2]catenanes 19' and 20', contain two disconnected rings since the photochemical reaction leads to decomplexation of the bipy chelate from the ruthenium(II) center. In a typical reaction, a degassed CH_2Cl_2 solution of 20^{2+} and NEt_4Cl was irradiated with visible light, at room temperature. The color of the solution rapidly changed from red (20^{2+}: $\lambda_{max} = 458$ nm) to purple (20': $\lambda_{max} = 561$ nm) and after a few minutes the reaction was complete. The recoordination reaction 20' → 20^{2+} was carried out by heating a solution of 20'. The quantum yield for the photochemical reaction 20^{2+} → 20' at 25 °C and $\lambda = 470$ nm (± 50 nm) can be very roughly estimated as 0.014 ± 0.005. One of the weak points of the present system is certainly the limited control over the shape of the photoproduct since the decoordinated ring can occupy several positions. It is hoped that, in the future, an additional tuneable interaction between the two rings of the present catenanes, 19' or 20' will allow better control over the geometry of the whole

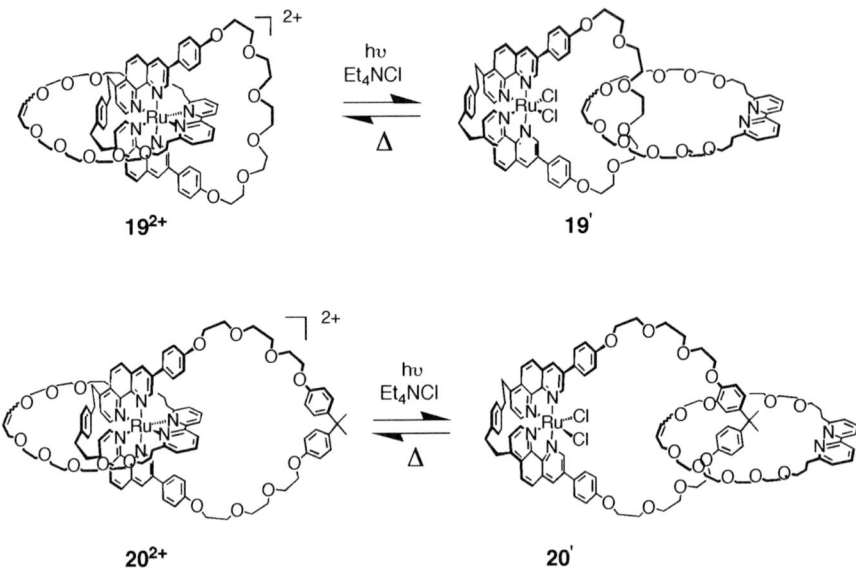

Fig. 28 Catenanes 19^{2+} or 20^{2+} undergo a complete rearrangement by visible light irradiation. The bipy-containing ring is efficiently decoordinated in the presence of Cl^-. By heating the photoproducts $19'$ or $20'$, the starting complexes 19^{2+} or 20^{2+} are quantitatively regenerated

system. In parallel, two-color machines will be elaborated, for which both motions will be driven by photonic signals operating at different wavelengths.

5
Conclusion and Prospective

The most important motivation for researchers in many groups, including ours, involved in the field of artificial molecular machines and motors, is certainly the synthetic challenge that the elaboration of such systems represents. It is indeed very challenging to reproduce some of the simplest functions of the natural biological motors (motor proteins, DNA polymerase, bacterial flagella, etc.) using synthetic molecular systems. It must nevertheless be kept in mind that the presently accessible molecular machines and motors are extremely primitive compared to the beautiful and exceedingly complex molecular machines of Nature. It must also be stressed that the study of molecular machines as single molecules also represents another very ambitious task. Until now, artificial systems have been almost exclusively been investigated as collections of molecules in solution or in films.

As far as practical applications are concerned, several possibilities can be explored although technological applications are probably not for tomorrow.

Information storage and processing at the molecular level is for the moment the most popular field of application since the spectacular reports of Heath, Stoddart and coworkers [124]. Other ambitious and futuristic practical outcomes could be considered such as the fabrication of "microrobots" or even "nanorobots" able to perform various functions: transport molecules or ions through a membrane, sort different molecules, or store energy, just to cite a few. In medicinal chemistry, it is conceivable that such devices can carry a given drug to a specific target where it is needed, open or close a gate or a valve that controls delivery of a drug from a microdevice, or act as a nanosyringe able to inject a given molecule inside a cell, among the many functions that nanomechanical devices should be able to fulfil in the future.

Acknowledgements We would like to thank all the very talented and enthusiastic researchers who participated in the work discussed in the present review article. Their names appear in the references. We would also like to thank the CNRS and the European Commission for their constant financial support.

References

1. Balzani V, Venturi M, Credi A (2003) Molecular devices and machines: a journey into the nanoworld. Wiley-VCH, Weinheim
2. Sauvage J-P (2001) Molecular machines and motors, Structure & Bonding. Springer, Berlin, Heidelberg, New York
3. Balzani V, Credi A, Raymo FM, Stoddart FJ (2000) Angew Chem Int Ed Eng 39:3348
4. Balzani V, Credi A, Venturi M (2003) Pure Appl Chem 75:541
5. Special issue on molecular machines (2001) Acc Chem Res 34:341
6. Feringa BL (2001) Molecular switches. Wiley-VCH, Weinheim
7. Fabbrizzi L, Licchelli M, Pallavicini P (1999) Acc Chem Res 32:846
8. Kelly TR, de Silva H, Silva RA (1999) Nature 401:150
9. Koumura N, Zijistra RWJ, van Delden RA, Harada N, Feringa BL (1999) Nature 401:152
10. Leigh DA, Wong JKY, Dehez F, Zerbetto F (2003) Nature 424:174
11. Jimenez-Molero MC, Dietrich-Buchecker C, Sauvage J-P (2003) Chem Comm 1613
12. Katz E, Lioubashevsky O, Willner I (2004) J Am Chem Soc 126:15520
13. Shipway AN, Willner I (2001) Acc Chem Res 34:421
14. Harada A (2001) Acc Chem Res 34:456
15. Schill G (1971) Catenanes, rotaxanes and knots. Academic, New-York
16. Sauvage J-P, Dietrich-Buchecker CO (eds) (1999) Molecular catenanes, rotaxanes and knots. Wiley-VCH, Weinheim
17. Vögtle F, Dünnwald T, Schmidt T (1996) Acc Chem Res 29:451
18. Fujita M (1999) Acc Chem Res 32:53
19. Sauvage J-P (ed) (1993) Special issue New J Chem 17
20. Hoshimo T, Miyauchi M, Kawaguchi Y, Yamaguchi H, Harada H (2000) J Am Chem Soc 122:9876
21. Kawaguchi Y, Harada H (2000) Org Lett 2:1353
22. Bogdan A, Vysotsky MO, Ikai T, Okamoto Y, Boehmer V (2004) Chem Eur J 10:3324
23. Stanier CA, O'Connell MJ, Cleg W, Anderson HL (2001) Chem Comm 493

24. Stanier CA, Alderman SJ, Claridge TDW, Anderson HL (2002) Angew Chem Int Ed Eng 41:1769
25. Wisner JA, Beer PD, Drew MGB, Sambrook MR (2002) J Am Chem Soc 124:12469
26. Chichak K, Walsh MC, Branda NR (2000) Chem Comm 847
27. Breault A, Hunter CA, Mayers PC (1999) Tetrahedron 55:5265
28. Dürr H, Bossmann S (2001) Acc Chem Res 34:905
29. Iwamoto H, Itoh K, Nagamiya H, Fukazawa Y (2003) Tetrahedron Lett 44:5773
30. Fitzmaurice D, Rao SN, Preece JA, Stoddart JF, Wenger S, Zaccheroni N (1999) Angew Chem Int Ed Eng 38:1147
31. Gibson HW, Yamaguchi N, Hamilton L, Jones JW (2002) J Am Chem Soc 124:4653
32. Samori P, Jäckel F, Ünsal Ö, Godt A, Rabe JP (2001) Chem Phys Chem 7:461
33. Hodge P, Monvisade P, Owen GJ, Heatley F, Pang Y (2000) New J Chem 24:703
34. Chang SY, Choi JS, Jeong KS (2001) Chem Eur J 7:2687
35. Fujimoto T, Nakamura A, Inoue Y, Sakata Y, Kaneda T (2001) Tetrahedron Lett 42:7987
36. Kim K (2002) Chem Soc Rev 31:96
37. Park KM, Kim SY, Heo J, Whang D, Sakamoto S, Yamaguchi K, Kim K (2002) J Am Chem Soc 124:2140
38. Korybut-Daszkiewicz B, Wieckowska A, Bilewicz R, Domagala S, Wozniak K (2001) J Am Chem Soc 123:9356
39. Chen L, Zhao X, Chen Y, Zhao CX, Jiang XK, Li ZT (2003) J Org Chem 68:2704
40. Davidson GJE, Loeb SJ (2003) Angew Chem Int Ed Eng 42:74
41. Coumans RGE, Elemans JAAW, Thordarson P, Nolte RJM, Rowan AE (2003) Angew Chem Int Ed Eng 42:650
42. Arduini A, Ferdani R, Pochini A, Secchi A, Ugozzoli F (2000) Angew Chem Int Ed Eng 39:3453
43. McArdle CP, Vittal JJ, Puddephatt RJ (2000) Angew Chem Int Ed Eng 39:3819
44. Udachim KA, Wilson LD, Ripmeester JA (2000) J Am Chem Soc 122:12375
45. Gunter MJ, Bampos N, Johnstone KD, Sanders JKM (2001) New J Chem 25:166
46. Ghosh P, Mermagen O, Schalley CA (2002) Chem Comm 2628
47. Andrievsky A, Ahuis F, Sessler JL, Vögtle F, Gudat D, Moini M (1998) J Am Chem Soc 120:9712
48. Belaissaoui A, Shimada S, Ohishi A, Tamaoki N (2003) Tetrahedron Lett 44:2307
49. Shukla R, Deetz MJ, Smith BD (2000) Chem Comm 2397
50. Smukste I, Smithrud DB (2003) J Org Chem 68:2547
51. Simone DL, Swager TM (2000) J Am Chem Soc 122:9300
52. MacLachlan MJ, Rose A, Swager TS (2001) J Am Chem Soc 123:9180
53. Watanabe N, Yagi T, Kihara N, Takata T (2002) Chem Comm 2720
54. Watanabe N, Kihara N, Furusho Y, Takata T, Araki Y, Ito O (2003) Angew Chem Int Ed Eng 42:681
55. Willner I, Pardo-Yissar V, Katz E, Ranjit KT (2001) J Electro Chem 497:172
56. Bissell RA, Córdova E, Kaifer AE, Stoddart JF (1994) Nature 369:133
57. Livoreil A, Dietrich-Buchecker CO, Sauvage J-P (1994) J Am Chem Soc 116:9399
58. Collin J-P, Gaviña P, Sauvage J-P (1996) Chem Comm 2005
59. Raehm L, Kern J-M, Sauvage J-P (1999) Chem Eur J 5:3310
60. Sauvage J-P (1998) Acc Chem Res 31:611
61. Badjic JD, Balzani V, Credi A, Silvi S, Stoddart JF (2004) Science 303:1845
62. Ballardini R, Balzani V, Gandolfi MT, Prodi L, Venturi M, Philp D, Ricketts HG, Stoddart JF (1993) Angew Chem Int Ed Eng 32:1301
63. Livoreil A, Sauvage J-P, Armaroli N, Balzani V, Flamigni L, Ventura B (1997) J Am

Chem Soc 119:12114
64. Armaroli N, Balzani V, Collin J-P, Gaviña P, Sauvage J-P, Ventura B (1999) J Am Chem Soc 121:4397
65. Brouwer AM, Frochot C, Gatti FG, Leigh DA, Mottier L, Paolucci F, Roffia S, Wurpel GWH (2001) Science 291:2124
66. Ashton PR, Ballardini R, Balzani V, Credi A, Dress KR, Ishow E, Kleverlaan CJ, Kocian O, Preece JA, Spencer N, Stoddart JF, Venturi M, Wenger S (2000) Chem Eur J 6:3558
67. Ashton PR, Ballardini R, Balzani V, Baxter I, Credi A, Fyfe MCT, Gandolfi MT, Gómez-López M, Martínez-Díaz M-V, Piersanti A, Spencer N, Stoddart JF, Venturi M, White AJP, Williams DJ (1996) J Am Chem Soc 120:11932
68. Balzani V, Credi A, Mattersteig G, Mattews OA, Raymo FM, Stoddart JF, Venturi M, White AJP, Williams DJ (2000) J Org Chem 65:1924
69. Murakami H, Kawabuchi A, Kotoo K, Kunitake M, Nakashima N (1997) J Am Chem Soc 119:7605
70. Balzani V, Credi A, Marchioni F, Stoddart JF (2001) Chem Comm 1860
71. Dietrich-Buchecker CO, Sauvage J-P, Kintzinger J-P (1983) Tetrahedron Lett 24:5095
72. Dietrich-Buchecker CO, Sauvage J-P, Kern J-M (1984) J Am Chem Soc 106:3043
73. Dietrich-Buchecker CO, Sauvage J-P (1990) Tetrahedron 46:503
74. Chambron J-C, Dietrich-Buchecker CO, Sauvage J-P (1996) In: Atwood JL, Davies JED, MacNicol DD, Vögtle F, Lehn J-M, Sauvage J-P, Hosseini MW (eds) Comprehensive supramolecular chemistry. Pergamon, Oxford 9:43
75. Dietrich-Buchecker CO, Sauvage J-P (1987) Chem Rev 87:795
76. Chambron J-C, Collin J-P, Heitz V, Jouvenot D, Kern J-M, Mobian P, Pomeranc D, Sauvage J-P (2004) Eur J Org Chem 1627
77. Pomeranc D, Jouvenot D, Chambron J-C, Collin J-P, Heitz V, Sauvage J-P (2003) Chem Eur J 9:4247
78. Mobian P, Kern J-M, Sauvage J-P (2004) Angew Chem Int Ed Eng 43:2392
79. Mobian P, Kern J-M, Sauvage J-P (2003) Helv Chim Acta 86:4195
80. Mobian P, Kern J-M, Sauvage J-P (2003) J Am Chem Soc 125:2016
81. Sauvage J-P, Weiss J (1985) J Am Chem Soc 107:6108
82. Dietrich-Buchecker CO, Khemiss A, Sauvage J-P (1986) Chem Comm 1376
83. Mitchell DK, Sauvage J-P (1988) Angew Chem Int Ed Eng 27:930
84. Chambron J-C, Heitz V, Sauvage J-P (1993) J Am Chem Soc 115:12384
85. Chambron J-C, Sauvage J-P (1998) Chem Eur J 4:1362
86. Blanco MJ, Chambron J-C, Heitz V, Sauvage J-P (2000) Org Lett 2:3051
87. Mohr B, Weck M, Sauvage J-P, Grubbs RH (1997) Angew Chem Int Ed Eng 36:1308
88. Rapenne G, Dietrich-Buchecker CO, Sauvage J-P (1999) J Am Chem Soc 121:994
89. Baumann F, Livoreil A, Kaim W, Sauvage J-P (1997) Chem Comm 35
90. Dietrich-Buchecker CO, Sauvage J-P, Kern J-M (1989) J Am Chem Soc 111:7791
91. Parthenopoulos DA, Rentzepis PM (1989) Science 245:843
92. Willner I, Blonder R, Dagan A (1994) J Am Chem Soc 116:3121
93. Irie M, Miyatake O, Uchida K (1992) J Am Chem Soc 114:8715
94. Gilat SL, Kawai SH, Lehn J-M (1993) Chem Comm 1439
95. Cardenas D, Livoreil A, Sauvage J-P (1996) J Am Chem Soc 118:11980
96. Schill G, Zollenkopf H (1969) Liebigs Ann Chem 721:53
97. Harrison IT, Harrison S (1967) J Am Chem Soc 89:5723
98. Ogino H (1981) J Am Chem Soc 103:1303
99. Bélohradsky M, Raymo FM, Stoddart JF (1996) Collect Czech Chem Commun 61:1
100. Gibson H, Bheda MC, Engen PT (1994) Prog Polym Sci 19:843

101. Chambron J-C, Harriman A, Heitz V, Sauvage J-P (1993) J Am Chem Soc 115:6109
102. Diederich F, Dietrich-Buchecker CO, Nierengarten J-F, Sauvage J-P (1995) Chem Comm 781
103. Zhu SS, Carroll PJ, Swager TM (1996) J Am Chem Soc 118:8713
104. Collin J-P, Gaviña P, Sauvage J-P (1997) New J Chem 21:525
105. Nicholson RS, Shain I (1964) Anal Chem 36:706
106. Poleschak I, Kern J-M, Sauvage J-P (2004) Chem Comm 474
107. Linke M, Chambron J-C, Heitz V, Sauvage J-P, Semetey V (1998) Chem Comm 2469
108. Linke M, Chambron J-C, Heitz V, Sauvage J-P, Semetey V (1997) J Am Chem Soc 119:11329
109. Jiménez C, Dietrich-Buchecker C, Sauvage J-P (2000) Angew Chem Int Ed Eng 39:3284
110. Jiménez C, Dietrich-Buchecker C, Sauvage J-P (2002) Chem Eur J 8:1456
111. Collin J-P, Laemmel A-C, Sauvage J-P (2001) New J Chem 25:22
112. Laemmel A-C, PhD thesis, University of Strasbourg
113. Adelt M, Devenney M, Meyer TJ, Thompson DW, Treadway JA (1998) Inorg Chem 37:2616
114. Van Houten J, Watts J (1978) Inorg Chem 17:3381
115. Suen HF, Wilson SW, Pomerantz M, Walsch JL (1989) Inorg Chem 28:786
116. Pinnick DV, Durham B (1984) Inorg Chem 23:1440
117. Durham B, Caspar JV, Nagle JK, Meyer TJ (1982) J Am Chem Soc 104:4803
118. Tachiyashiki S, Mizumachi K (1994) Coord Chem Rev 132:113
119. Pomeranc D, Chambron J-C, Heitz V, Sauvage J-P (2001) CR Acad Sci 4:197
120. Pomeranc D, Chambron J-C, Heitz V, Sauvage J-P (2001) J Am Chem Soc 123:12215
121. Jouvenot D, Koizumi M, Collin J-P, Sauvage J-P (2005) Eur J Inorg Chem 1850
122. Hayoz P, von Zelewsky A, Stoeckli-Evans H (1993) J Am Chem Soc 115:51111
123. Dietrich-Buchecker CO, Rapenne G, Sauvage J-P (1999) Coord Chem Rev 185–186:167
124. Collier CP, Mattersteig G, Wong EW, Luo Y, Beverly K, Sampaio J, Raymo FM, Stoddart JF, Heath JR (2000) Science 289:1172

Altitudinal Surface-Mounted Molecular Rotors

Thomas F. Magnera · Josef Michl (✉)

Department of Chemistry and Biochemistry, University of Colorado,
Boulder, CO 80309-0215, USA
magnera@eefus.colorado.edu, michl@eefus.colorado.edu

1	Introduction	64
1.1	Molecular Rotors	64
1.2	Surface-Mounted Rotors	65
1.3	Altitudinal Rotors	66
1.4	Organization of the Project	68
2	Rotor Design and Synthesis	69
2.1	Covalent Synthesis	70
2.1.1	Metal Sandwich Stands	70
2.1.2	Triptycene Stands	73
2.2	Synthesis by Self-Assembly	74
3	Surface Mounting	77
3.1	Adsorption Monitoring	78
3.2	Surface Footprint	79
3.3	Surface Conformation	81
4	Rotor Performance	85
4.1	Barrier Height Imaging	85
4.2	Computation of Demands on the Driving Force	88
4.2.1	Low Temperature	90
4.2.2	Elevated Temperatures	93
5	Summary and Outlook	94
	References	95

Abstract After a general description of some potential applications of artificial altitudinal surface-mounted molecular rotors, the synthesis, surface mounting, and characterization of the first examples of such rotors on a gold surface are described. Molecular dynamics calculations are used to discuss how a unidirectional rotation can be induced by an electric field oscillating normal to the surface as a function of its amplitude and frequency, as well as temperature.

Keywords Molecular rotor · Altitudinal · Surface-mounted · Barrier height imaging · Molecular dynamics · Single-molecule parametric oscillator

1
Introduction

1.1
Molecular Rotors

Most macroscopic machines contain rotating parts. Nature has found it expedient to build them from proteins [1, 2], and it is likely that artificial molecular-size machines [3, 4] will need them, too. Contemplation of intramolecular rotation seems to hold special fascination to many chemists, and attempts to design, synthesize, and examine molecules with unique rotational properties started a long time ago. Quite possibly, it is the discovery of atropisomerism in substituted biphenyls [5] or the discovery of a barrier to internal rotation in ethane [6] that ought to be viewed as opening the era of molecular rotors. This field then witnessed a spectacular increase in sophistication [7] through the successful demonstration of molecular gears [8] to the more recent synthesis of molecular rotors that can be slowly unidirectionally driven by controlled irradiation [9, 10] or with chemical reactions [11].

It has become common to apply the term molecular rotor to the whole molecule, not just the rotating part. We follow this usage and have adopted the expression rotator when referring to the part of the molecule that turns relative to the rest. Instead of molecular rotor, some authors would use the term molecular motor, but our personal preference is to reserve the term motor for devices that actually perform useful work and do not merely idle and convert other forms of energy into heat.

As the examples of ethane and biphenyl amply demonstrate, it is not easy to distinguish the rotator from the stator inside a molecular rotor. When a part of a free molecule turns, the rest turns in the opposite sense so as to conserve angular momentum, and the molecule ends up oriented in space in a way that depends on which particular internal motion has taken place [12]. If the internal motion is a rigid rotation around a single axis, it is the relative size of the moments of inertia of the two mutually rotating parts relative to the common rotation axis that determines how much of a rotation each part performs. An exception to the claim that it is hard to distinguish the rotator from the stator in a molecule are molecular rotors that are firmly anchored to a macroscopic object, whose moment of inertia is virtually infinite on the molecular scale. Among such mounted molecular rotors, surface-mounted rotors have been of particular interest to us since we first started to contemplate the synthesis of rotating molecular-size objects [13] as a part of a larger project dubbed "molecular Tinkertoys" [14], assembly of molecule-sized objects from a set of rods and connectors [15, 16].

1.2
Surface-Mounted Rotors

There are other reasons why we found the fabrication of surface-mounted rotors more appealing than work with rotors freely floating in a solution, or rotors mounted inside a solid. If the rotors are to become a part of a coherent larger assembly and if they are to communicate with other moving parts, it will be a considerable advantage if their location and orientation are well defined. In our view, even the simplest aggregate of identical parts working together will be easier to produce on a surface. Experimental differentiation of internal from external molecular motion is also facilitated. Of course, there is a price to pay, in that the characterization of surface-mounted rotors requires the demanding techniques of surface science rather than the more common tools of solution chemistry, and it took some effort to introduce these techniques to our laboratory. The tiny amounts with which one can work on a surface impose other obvious limitations, but also offer advantages. On the one hand, if one were to use surface-mounted molecular rotors in optoelectronics, for example, one might need many surface layers to achieve a useful effect. On the other hand, only small amounts of new materials need to be synthesized, as a 1-mg sample will keep a surface scientist happy for many months.

We considered anchoring the molecular rotors inside a solid, which would make the work easier in many respects. However, we ultimately wanted to organize the rotors into controlled arrays, and, right or wrong, crystal engineering in three dimensions looked harder than surface assembly in two. For further manipulation, surface-mounted rotors would have the advantage of offering access to all components at all times from the third direction. After all, there is a reason why computer chip architecture is largely two-dimensional, and we thought that it might be possible later to use epitaxy to grow multilayer structures based on active elements carried by a grid that is periodic only in two dimensions and aperiodic in the third, a "designer solid" [17]. Molecular dynamics simulations of dipolar molecular rotors that we performed for grid-mounted rotors [18] and subsequently for rotors mounted between grid layers [19] led us to expect large friction for a dipolar rotor located inside a crystal. This did not bode well for some of the applications we had in mind, such as examination of rotational excitation propagating through a rotor array. According to these calculations, even in the absence of mechanical interference, the interaction of the moving charges with those in the surroundings causes rapid transfer of energy and angular momentum to the bulk of the solid. Even on a surface, the friction was calculated to be disturbingly large.

Finally, there was the issue of novelty. Others were doing beautiful work preparing and studying molecular rotors freely floating in solution, and many solids exhibiting internal rotation were also already known, but artificial surface-mounted rotors appeared entirely new. Our initial attempts were in the direction of azimuthal rotors, with an axle mounted perpendicular to a surface [20], and they are continuing in the hope that we will learn how to assemble regular arrays of such rotors for the study of collective phenomena. Some advances in that direction have been reported [21, 22]. Presently, however, we shall describe related work on altitudinal rotors, which have an axle mounted parallel to a surface.

1.3
Altitudinal Rotors

It seemed to us initially that altitudinal rotors would be harder to produce, since the axle needs to be held by two stands tall enough to permit the rotator to clear the surface. Indeed, the synthesis is more complicated, but the surface mounting has turned out to be easier and sturdier. Rotors of this type look a little like paddle wheels, and when the number of paddles on the axle exceeds two, they evoke images of a water mill on a creek. Then, it is reasonable to ask whether one could fabricate surfaces capable of interacting with parallel flow of a liquid or a gas. The gap between the macroscopic and the nanoscopic world is always larger than it seems to be, and it is not at all clear that such an interaction will be discernible above the thermal noise of Brownian motion. Conceivably, if the rotors can be designed in a way that makes the coupling of their rotation with the flow strong enough, their presence will affect the friction experienced by the flow. Further, if the axes of the rotors could be aligned parallel to each other, such friction might become anisotropic. To make this happen, we need not only the ability to synthesize rotors whose rotation couples to fluid flow strongly enough, but also the ability to align them on the surface in a preordained fashion.

Demonstrating that surface-mounted molecular rotors can be driven mechanically would be a first step and a fair accomplishment in itself. Ultimately, however, it would be much more interesting to design rotors that can be driven at high speed by absorption of light pulses, or perhaps by an alternating electric field, and that actively pump a very thin layer of fluid parallel to the surface on which they are mounted. Active transport could be considered useful work, and these would then indeed be molecular motors by anyone's definition. In order to make this advance, we would most likely need to learn how to drive the rotors at microwave frequencies and we would definitely need to learn how to orient them on the surface, not merely how to align them. If half of them pumped in one direction and half in the opposite direction randomly, the assembly would be useless as a pump, even though it might still serve as a stirrer. We do not know whether we will be able to speed up light-driven molecular rotors, which currently operate many orders of magnitude more slowly than we require, and to design their completely controlled self-assembly on a surface into a mosaic array in which each tile knows exactly where all four directions of the wind-rose are, but it surely is an interesting challenge.

However, just because altitudinal rotors can be made to look like paddle wheels, nanofluidics is not the only direction towards which the development of surface-mounted altitudinal rotors can be oriented. Even without any rotors, developing a capability for completely controlled surface tiling would serve many other purposes. An example is molecular electronics, a fascinating field that attracts much attention [23], although useful single-molecule electronics does not exist presently and possibly never will. All manner of single-molecule electronic devices built around molecules that can change their conformation by rotation can be imagined and several have been proposed [24, 25].

Much closer to reality is the other meaning of the expression molecular electronics, which refers to the use of macroscopic amounts of molecularly engineered materials. Here, too, it is easy to envisage numerous interesting applications, even for altitudinal rotors that only have two paddles. For instance, a field of rotors that carry a sufficiently large dipole could be arranged in a rectangular lattice with all axles parallel. The array might then have two macroscopically dipolar stable states and serve as a storage device. In one state, all rotors would have their dipoles pointing to the east; in the other, all would be pointing to the west. Switching between the two states should be possible by raising the temperature momentarily in the presence of an orienting electric field.

Other applications can be envisaged in optoelectronics and we provide a single example. Consider a layer of altitudinal rotors whose rotator consists of a large flat π-electron system, endowed with a large in-plane polarizability in one direction perpendicular to the axis and a small out-of plane polarizability in the other. The system can be chosen to be a chromophore with $\pi\pi^*$ absorption in the visible, and a transition moment directed perpendicular to the axis of the rotor. The rotors need to be designed in such a way that in their ground state, they will arrange their planes and transition moments perpendicular to the surface. The layer will then be transparent in the visible. If necessary, the rotors can be encouraged to adopt this ground-state arrangement by the presence of a dipole parallel to the transition moment; either a ferroelectric or an antiferroelectric coupling would be satisfactory. The electric field of a strong laser pulse will act on the anisotropic polarizability and turn the rotor blades, and thus the visible transition moments, parallel to the surface. Once the rotors are turned, the layer will absorb further incident light, and a sufficient number of layers will absorb practically all of it. If a device of this type turns out to be feasible, it could be used for extremely fast protection against strong laser pulses.

1.4
Organization of the Project

After this brief motivation of our efforts in the area of molecular rotors, and before a description of what has been accomplished so far, we provide an outline of the way in which our molecular rotor project is organized. This chapter is organized along similar lines.

The first step in the production of a device based on a collection of surface-mounted molecular rotors is computer-aided design of a suitable structure, a molecular dynamics simulation of its operation, and its actual synthesis in the laboratory. All these need to be done simultaneously and interactively to assure both synthetic feasibility and proper function. It is just as easy to design intricate structures on the computer screen whose synthesis would require a decade of effort as it is to synthesize randomly chosen molecules that look promising but then fail to perform. Standard molecular dynamics programs are not suitable for such modeling, since they do not easily permit the use of outside time-dependent fields, of gas and liquid flow, of atoms of less common elements, such as gold, nor do they allow the incorporation of metal surfaces in the structures to be examined. We therefore developed our own molecular dynamics program TINK [18], starting with code kindly provided by Prof. Elber. It uses the UFF force field, which is very approximate, but is available for all elements of the periodic system [26], and partial atomic charges determined by charge equilibration [27]. The program permits the introduction of time-dependent outside electric fields, entry and departure of atoms as needed in flow simulations, macroscopic motion of surfaces against each other, use of periodic boundary conditions, and work with image charges inside a conductor.

We realize that the approximations involved in the calculations, in particular the neglect of electronic polarizability, are severe. Although the results do not guarantee that the rotors will perform as expected, rough guidance is still infinitely better than none. Once the rotors have been synthesized, their ordinary solution properties, such as barriers to rotation, can be checked against expectations and severe design flaws can be eliminated at this stage.

The next step is the mounting of the rotors on a suitably prepared surface. Since altitudinal rotor molecules tend to be large, their volatility at accessible temperatures is generally insufficient for adsorption from the gas phase, and we use adsorption from solution. This needs to be monitored to assure a submonolayer or at most monolayer coverage.

Characterization of the orientation of the adsorbed molecules on the surface follows, and it is crucial. Careful "PowerPoint design" chemistry does not yet guarantee that in reality the molecule adopts the required conformation on a surface.

Finally, it is time to demonstrate that the altitudinal rotor can actually turn on the surface as desired and is not blocked by contaminants, protruding surface asperities, or disabled for some other reason. This is the most critical part of the whole process, and so far we have managed to complete it with only one of our altitudinal molecular rotors.

In the future, we need to design altitudinal rotor structures that will permit us to proceed further. In particular, we need to endow them with the ability to self-assemble on the surface into regular two-dimensional arrays, and with the ability to respond to energy input by rapid rotation.

2
Rotor Design and Synthesis

The simplest altitudinal rotor has three essential components: two identical stands with affinity to a surface, an axle, and a rotator (Fig. 1). The height of the stands needs to be matched to the size of the rotator to make sure that the rotator clears the surface during its rotatory motion. The axle is a molecular rod and will therefore be far more flexible than would be desirable [28], requiring a large clearance allowance. The stands, too, should be as rigid as possible, and so should be their attachment to the surface. Chemisorption, as opposed to mere physisorption, would appear essential, and a wide stocky stump would appear better than a thin pole. For control experiments, the rotator should be available in a non-polar as well as a polar version.

Our choice of actual rotor structures has been influenced by prior experience with other projects. Previous work with sandwich complexes of cobalt with a pentafunctionalized cyclopentadienyl as one deck and tetraarylcy-

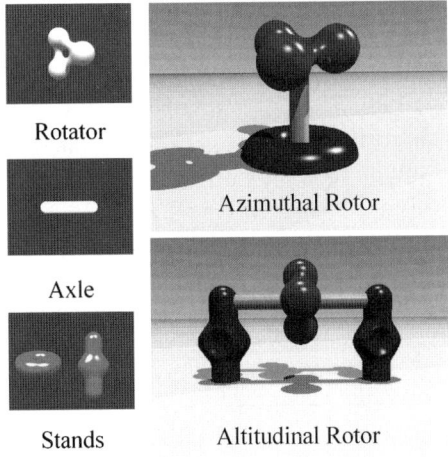

Fig. 1 Surface-mounted molecular rotors and their components

clobutadiene as the other [29] prompted us to choose a similar structure as very stocky stands for our first generation altitudinal rotor. This is the molecular rotor for which we have obtained the most results and the bulk of this chapter will deal almost exclusively with it. On the basis of calculations, we believe that this first generation rotor has a barrier to rotation of several kcal/mol when mounted on a surface, and this has advantages and disadvantages. In the design of the second generation of altitudinal rotors, we attempted to minimize this barrier. In doing so, it is important to recognize that the potential energy profile for rotation will depend not only on the molecular rotor itself, but will be a function of the nature of the substrate. The largest difference will be between conducting and insulating surfaces, since interaction of the rotating dipole with image charges, present in the former, will be absent in the latter.

Rotors of the first and second generations were quite laborious to synthesize, and we have also examined the notion that it might be possible to prepare suitable rotors in just a few steps, using the principles of self-assembly. Recent progress in this direction will be described.

2.1
Covalent Synthesis

From the perspective of retrosynthesis, a natural first disconnection of an altitudinal rotor is common to all of our rotors. It takes advantage of their symmetry and splits them into two identical stands and an axle-rotator unit. The surface-active functionalities attached to the stands are likely to be highly reactive and it is probably best to introduce them last.

2.1.1
Metal Sandwich Stands

The synthesis of the polar and non-polar versions our first altitudinal rotor is shown in Scheme 1 [30]. The polar rotator is a 9,10-dihydrophenanthrene substituted by four fluorine atoms in the central ring. It is calculated to have a dipole moment of about 3.7 D directed along the axis of two-fold symmetry. The non-polar rotator is 4,5,9,10-tetrahydropyrene, whose D_2 symmetry precludes a dipole moment. The sequence of synthetic steps used has ample precedent and requires no explanation. The most questionable step in the synthesis appeared to be the ten-fold mercuration, which however proceeded in good yield and turned out to be no problem at all.

The anticipated advantage of using the $-Hg-O_2C-CF_3$ substituent at the bottom of the stands was its versatility. It reacts very easily with almost any thiol RSH with displacement of the trifluoroacetate anion and formation of the mercury thiolate $-Hg-SR$, permitting the facile introduction of a wide variety of residues R with desired surface affinity. We have chosen two, based

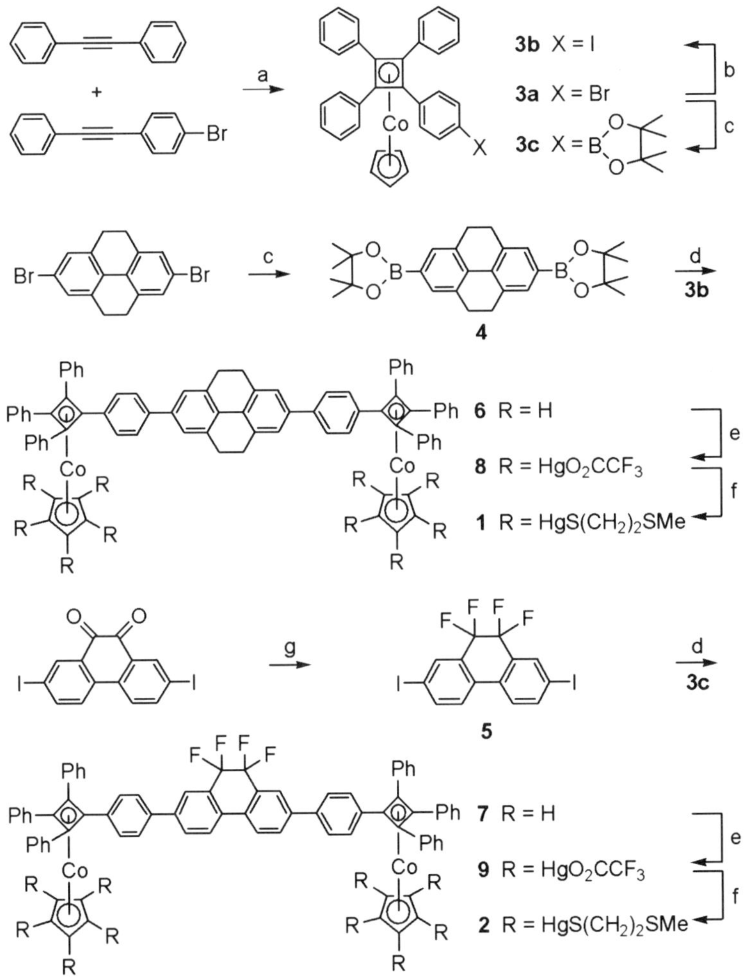

Key: a) CpCo(CO)$_2$; b) t-BuLi, I$_2$; c) bis(pinacolato)diboron, PdCl$_2$(dppf), KOAc; d) Pd black, KF; e) Hg(O$_2$CCF$_3$)$_2$; f) HS(CH$_2$)$_2$SMe; g) F$_3$SNEt$_2$.

Scheme 1 Synthesis of the first-generation altitudinal rotor: $\frac{1}{m}$, non-polar; $\frac{2}{m}$, polar

on prior work in our laboratory. Good experience with the adhesion of dialkyl sulfide containing chains to gold [31] prompted us to use the thiol CH$_3$SCH$_2$CH$_2$SH for work on gold surfaces, and good experience with the attachment of molecules to fused quartz surfaces using the trialkoxysilane moiety [20] prompted us to use the thiol (C$_2$H$_5$O)$_3$SiCH$_2$CH$_2$CH$_2$SH for work on quartz. As we shall see below, we found excellent adhesion in both cases, although it was somewhat of a shock to realize that for adhesion on gold, the initial –Hg–O$_2$C–CF$_3$ substituent is entirely adequate [32]. This discovery may have consequences in other areas of nanoscience where attachment of molecules to surfaces is critical, since attachment through a metal atom may provide enhanced electrical conductivity.

Although in Scheme 1 the dipolar and non-polar rotors are each drawn as a single species, they consist of mixtures of conformers. For each rotor, there are three pairs of enantiomers that differ in the helicity at the tetraarylcyclobutadiene decks of the two stands (P or M on each), and in the helicity of the rotator, whose two benzene rings are slightly twisted relative to each other. Placing the helicity symbol for the rotator in the center and the two symbols for the stands at each end, we end up with six conformer labels, PMP, PPM, PPP, MPM, MMP, and MMM. The first three are shown in Fig. 2. The interconversion of the six species involves only rotations around single bonds and is very facile, such that at room temperature all relevant NMR signals are averaged. In the dipolar rotor, cooling to $-100\,°C$ slowed down only one of the P-M interconversions sufficiently for analysis by dynamic ^{19}F NMR, and revealed an activation energy of 6.3 kcal/mol. The structural change associated with this barrier is the interconversion of pseudoequatorial and pseudoaxial fluorine substituents in the central ring of the rotator as it flips between its P and M enantiomeric forms. This was established by a study of 9,9,10,10-tetrafluorodihydrophenanthrene itself, which yielded a nearly identical barrier of 6.7 kcal/mol, and it also agrees with a B3LYP/6-31G* calculation for the latter, which provided a value of 6.2 kcal/mol. The barriers for the P-M interconversion of the stands and for the rotation of the rotator around the axle are apparently much smaller and their effects are not detectable in the NMR spectra down to $-100\,°C$. The rotational barrier for the turning of the rotator calculated by the B-LYP/SV(P) (ECP for Hg) method

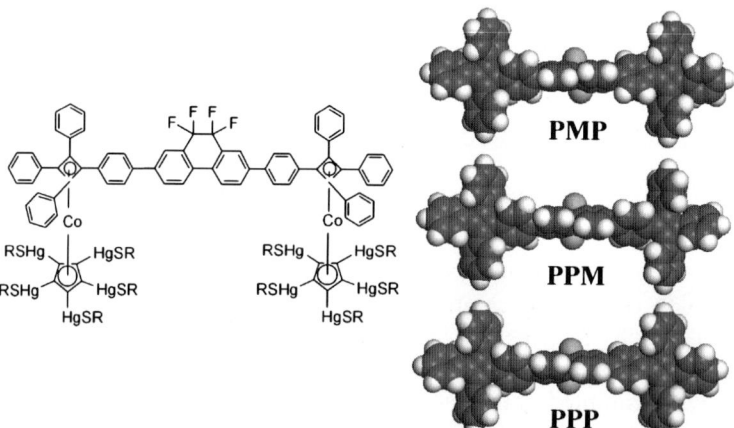

Fig. 2 Members of three pairs of enantiomeric conformers of the dipolar altitudinal rotor 2 (Scheme 1)

is 3 kcal/mol, comparable with the rotational barrier in biphenyl. Clearly, in the free rotor, the rotator flips fast at room temperature. Below, we shall ask whether this is still true when the rotor is mounted on a surface, and we shall see that the existence of small rotational barriers is expected to lead to some very interesting consequences.

2.1.2
Triptycene Stands

We were also interested in preparing an altitudinal rotor that would have no or little intrinsic barrier, and our synthetic approach to a second generation rotor is summarized in Scheme 2 [33]. An obvious way to provide the rotor axle with effectively cylindrical symmetry and thus to reduce intrinsic barriers to rotation is to insert triple bonds between the rotator and the stands. Triple bonds are, however, not compatible with the mercuration step used as one of the last reactions in the synthesis of the first generation rotor (Scheme 1), and we decided to use different stands. The second-generation triptycene-based stands shown in Scheme 2 have the advantage of placing the axle higher above the surface and therefore permitting the use of larger rotators. We have chosen triptycene for a non-polar rotator and difluorotrip-

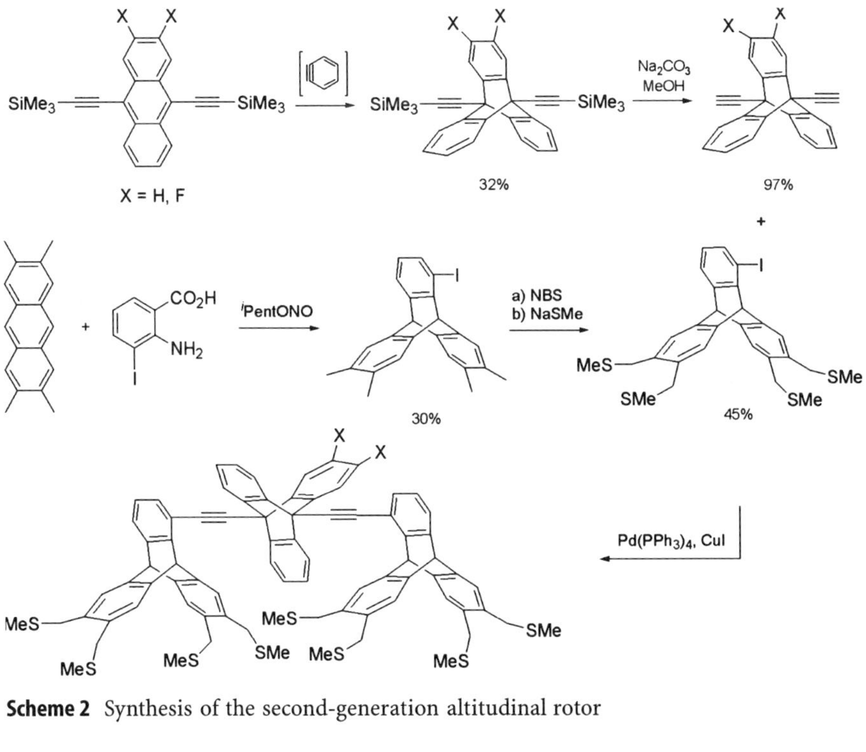

Scheme 2 Synthesis of the second-generation altitudinal rotor

tycene for a polar one, with an anticipated dipole moment of about 3 D. The triptycene framework was chosen since we wanted to gain experience with multi-paddled wheels for future work on flow-driven rotors.

The synthetic steps chosen (Scheme 2) again have ample precedent and require no comment. We originally hoped that the eight sulfur atoms of the methyl benzyl thioether type will provide adequate adhesion to the gold surface, and indeed, these rotor molecules clearly adsorb. Unfortunately, initial scanning electron microscopy (STM) images suggest that the adsorption strength may be inadequate, because the molecules tend to be swept aside by the tip. Perhaps the strength of the dialkyl sulfide–gold surface interaction is insufficient, but more likely and more ominously, perhaps some of the benzylic C – S bonds are cleaved by the gold surface. The latter possibility is suggested by some partially conflicting prior reports [34–36] of dialkyl sulfide C – S bond cleavage on a gold surface and by an observation that we made only after the synthesis of the second generation rotor had been completed, namely that the C – S bonds of *tert*-butyl alkyl sulfides are cleaved on a gold surface [32]. Although this observation still requires additional study before firm conclusions can be drawn, the close similarity in the reactivity of *tert*-butyl and benzyl groups now makes us anticipate that the adhesive functionalities on the second generation rotor (Scheme 2) will ultimately have to be modified. This illustrates one of the major difficulties of working with surface-mounted rotors: much less is known about the adsorption of various functionalities on metal surfaces than one might expect.

2.2
Synthesis by Self-Assembly

Metal-ion mediated self-assembly of coordination ligands [37–43] is a well-established rapid route to large molecules and it fits well with the original Tinkertoy philosophy [15, 16], although in the latter, covalent synthesis was envisaged in order to assure sturdiness of the assembled objects. Nevertheless, if the strength of the metal-ligand bonds is sufficient, there is no reason not to use them, provided that the assembled structure is electroneutral and not ionic. We believe that rotors with a net charge are less suitable for our purposes since the response of their dipolar rotators to outside electric fields might be overwhelmed by the response of their charged ionic parts, and especially, by the response of their possibly quite mobile counterions.

In spite of the appeal of its amazing conceptual simplicity, non-covalent synthesis of large molecular rotors by self-assembly faces considerable obstacles. First, only a limited amount of information is available on generally applicable synthetic strategies for reaching predefined targets, and components often assemble in unexpected ways. Second, almost all the squares, rectangles, cubes, prisms, and other complex geometrical objects that have been assembled using metal-mediated assembly were charged. This is understandable, since avoidance of defects requires thermodynamic control of the assembly process, and ligand exchange is generally much easier on metal ions than on electroneutral metal atoms. Third, it is presently relatively easy to assemble a cube or a prism with all symmetry-equivalent edges made of one kind of rod, but little is known about differentiating the edges. Perhaps each edge can carry all the functionalities required, e.g. affinity for a surface as well as a functional rotator, but it would seem better if otherwise equivalent edges did not need to all be the same, allowing one or two to carry rotators while others adhere to a surface.

We started our venture into the territory of molecular rotor self-assembly by (i) accepting charged structures as a necessary temporary evil; (ii) accepting the equivalence of all rods as another necessary temporary evil; and (iii) identifying right prisms as target structures. Such prisms consist of two parallel equivalent polygons of n vertices connected by n equivalent parallel straight lines perpendicular to the polygons. At least one of these parallel lines is to carry a rotator, and one of the rectangular faces defined by two neighboring parallel lines and one edge from each of the two polygons is to adsorb to a surface (Fig. 3). The ways in which the temporary compromises in items (i) and (ii) are to be removed later are under examination now. An obvious choice for charge removal is its compensation with an opposite charge located very closely nearby, i.e., rigid incorporation of the counterion into the rotor structure. If this is done, it would be best for the resulting dipole to be as small as possible and to lie in the axis of the rotator-carrying rod, since it would then be perpendicular to the dipole carried by the rotator, and it would interfere the least with the outside field used to communicate with the dipole of the rotator. An even better procedure would be a replacement of components attached by metal-ligand bonds with equivalent components attached by covalent metal-carbon bonds while preserving the self-assembled structure, but such conversions are not currently known. A possible way to remove rod equivalence would be to tether inequivalent rods in appropriate combi-

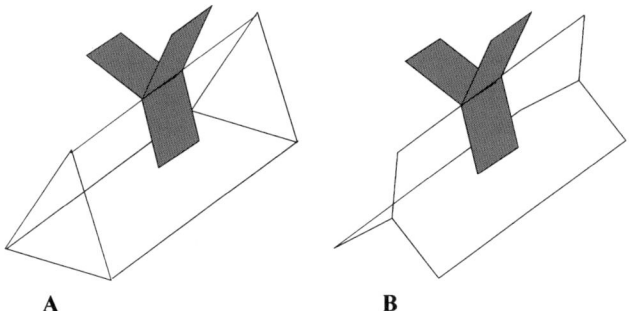

Fig. 3 A surface-mounted trigonal prismatic molecular rotor. **A**: prism defined by molecular rods along all nine edges; **B**: prism defined by a combination of molecular rods along three edges and star connectors along two faces

nations before assembly begins. If all-covalent neutral prisms can be made, they might be sufficiently stable for chromatographic separation, permitting the use of rod mixtures in the self- assembly step.

Maximal mechanical strength would be built in if all $3n$ edges of the prism were represented by molecular rods (Fig. 3A). This seemed too hard as an initial target, and we settled for prisms in which the two polygons are defined by their faces and only the n parallel edges are represented by molecular rods (Fig. 3B). This requires assembly from two equivalents of face-defining star connectors with n arms and n equivalents of edge-defining rods. The metal atoms that connect the arms of the star connectors with the rod termini at right angles are to be located at the vertices of the prism. Since, for the time being, we have chosen all the rods to be equal, they all must carry rotators at their centers (in Fig. 3, only one of the rods is shown with a rotator). Surface affinity could also be carried by all the rods, or it could be carried by other ligands on the $2n$ metal atoms. In the latter case, it would need to be directed in such a way that only the rectangular faces and not the polygon have affinity to the surface, otherwise the rotor may end up being azimuthal instead of altitudinal.

The simplest approach would be to provide the termini of the connector arms and of the rods with reversible affinity for a suitable metal ion that carries two reactive sites in a *cis* disposition. The other sites of the metal ion would be blocked by permanent ligands, two if a square planar complex is used, and four if an octahedral complex is used. The self-assembly would then be from $2n$ equivalents of metal ions in addition to the two equivalents of star connectors and n equivalents of rods.

We were concerned that this strategy would leave too much room for error, in that the metal ion might be tempted to connect two rods to each other or two star connectors to each other, instead of only connecting rods to connectors. Since we do not need to have all newly produced bonds form reversibly, only the rods attached to connector arms reversibly, we decided to attach the metal ions covalently and irreversibly to each of the rod termini. This had to be done in a way that directs the remaining reactive metal site to bind laterally, in a direction perpendicular to the rod axis.

The alternative would have been to attach the metal ions irreversibly to the arms termini of the star connector. Although quite feasible, this appeared less convenient given the star connectors already available in our laboratory [44]. Also, to assure minimal interference from the resulting dipole by orienting it properly, we ultimately plan to incorporate two counterions into the two termini of each rod and not one into each of the n arms of the star connector. The counterion is to be as close to the metal ion as possible, and it seemed easier to build it into the functionality that makes the covalent bond than into the one that makes the coordination bond. This provided another argument in favor of covalent attachment of the metal ion to the rod.

We started the project with square planar metal complexes of metal ions, carrying (i) a double phosphine ligand intended to block two *cis* positions; (ii)

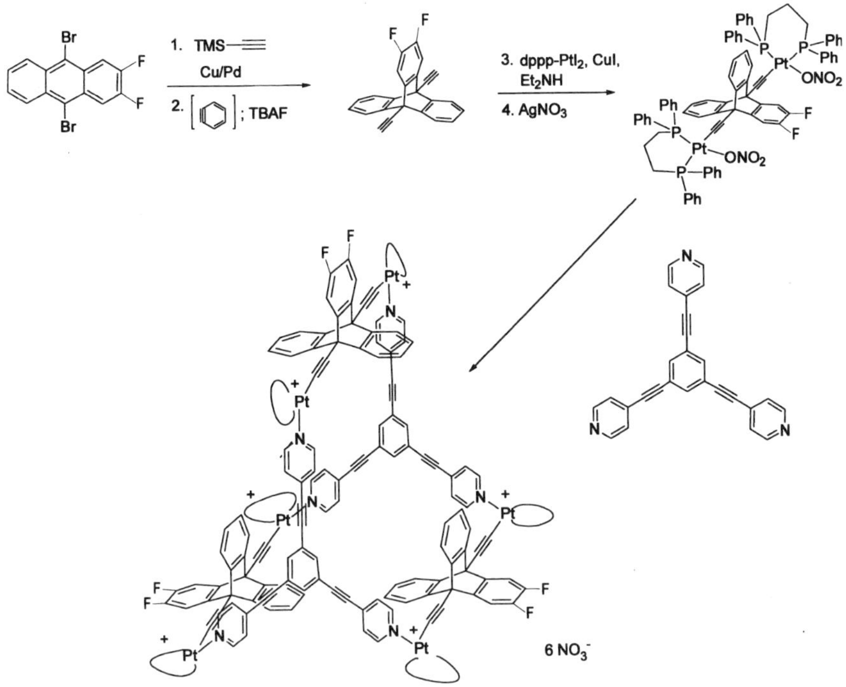

Scheme 3 Synthesis of a self-assembled altitudinal rotor

a good leaving group; and (iii) a covalent metal-carbon bond to the rod terminus. Pyridines represented the termini of the arms of the star connector. After some experimentation with palladium, we settled on platinum as a choice that provides sturdier structures. We first successfully tested the assembly of a rectangle [45], which is conceptually the simplest of prisms ("digonal"), and then tested the assembly of a trigonal and a tetragonal (square) prism [46]. These structures contained only simple molecular rods and no rotators.

Scheme 3 shows a more recent accomplishment, with dipolar rotators incorporated into three parallel edges of a trigonal prism [47]. The simplicity of the synthesis is striking, but of course it was preceded by a fair amount of preliminary effort [45, 46], and the two ultimate steps discussed above still remain to be done: charge neutralization and introduction of surface affinity, possibly with a removal of the two unneeded rotators.

3
Surface Mounting

Once a rotor has been synthesized and characterized in solution, it needs to be mounted on a suitable surface to complete the fabrication. In the remainder of this chapter, we shall focus attention on the non-polar (1) and polar (2) altitudinal rotors introduced in Scheme 1. Most of the work was done with rotors mounted on the surface of gold.

3.1
Adsorption Monitoring

Given the size of the rotor molecules, adsorption onto surfaces needs to be done from solution and not from the gas phase. We have controlled the rate of the process by a choice of concentration and exposure time and directed it to the formation of submonomolecular layers. This required a monitoring of the average thickness of the layer deposited.

On a gold surface, ellipsometry showed a steady increase of this thickness up to a complete monolayer coverage, at which point the growth slowed down substantially. This was confirmed by STM imaging, which showed an initial adsorption of oval-shaped images attributed to individual molecules, followed by the appearance of aggregates and finally by complete coverage. For further studies, we used samples that have relatively low coverage, up to about 1/4, in order to avoid rotor-rotor interference [30].

The strength of the adhesion is excellent. The adsorbed rotor molecules cannot be removed by rinsing with solvents and they are not pushed around easily with an STM tip. This was not a surprise, given the twenty sulfur atoms at the bottom of each rotor.

In order to assure that the chemical nature of the deposit indeed corresponds to the desired rotor and not to an adventitious contaminant, we obtained X-ray photoelectron spectra (XPS) on the very same samples. An example of an XPS of the dipolar rotor on gold is shown in Fig. 4. The characteristic doublet peaks of cobalt and mercury are clearly seen, as is the peak of fluorine. There are two partially overlapping doublets of sulfur, attributed to the dialkyl sulfide at higher binding energies and the mercury thiolate at lower energies. A similar XPS was obtained for a surface layer of a simple stand, without the rotor, except that the fluorine peak was absent, as it should be [30].

Exposure of the gold samples to air causes relatively rapid oxidation of the sulfur atoms. After several days, XPS signals reveal a complex process in which the sulfur atoms go through several oxidation states, and are ultimately converted to sulfonates. Unlike the original molecules, sulfonates would no longer be expected to adhere well to the gold surface. Most remarkably, some of the rotor molecules themselves remain on the surface and continue to be observable by STM. This made us realize that the sulfur-containing "tentacles" are actually not needed and that the ten mercury atoms apparently suffice to attach the rotor to gold quite adequately. We have subsequently performed experiments with rotors attached merely through their $-Hg-O_2CCF_3$ substituents [32]. According to XPS, the trifluoroacetate group

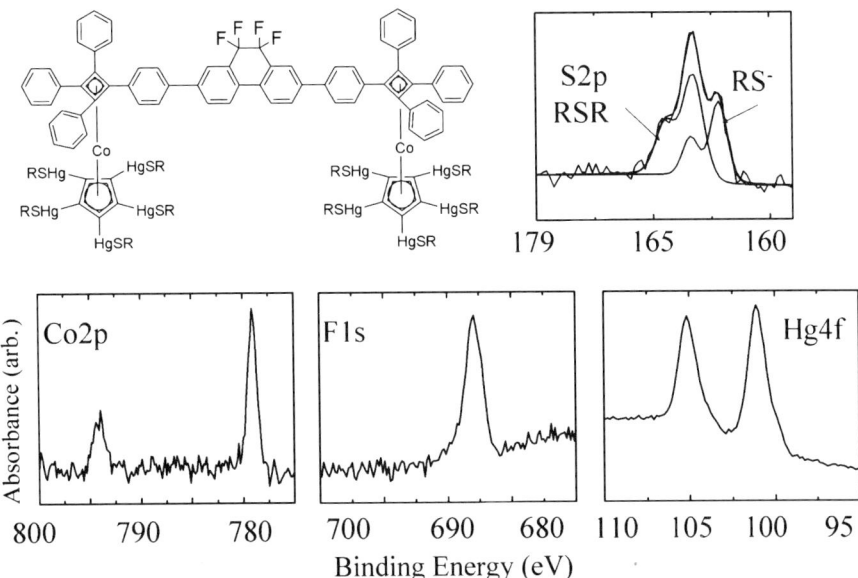

Fig. 4 X-ray photoelectron spectrum of dipolar rotor 2 on Au(111), **R** = HgSCH$_2$CH$_2$SCH$_3$

is lost in the adsorption process, and we suspect that the attachment to gold is through a nominally positively charged mercury cation, –Hg$^+$–Au(metal). This is compatible with the observed shift of the XPS signal of mercury to lower binding energies if the metal substrate feeds sufficient electron density to the mercury ion.

3.2
Surface Footprint

Adsorbing the rotor molecules on the surface and making sure that they are present at the required density represents only the first half of the surface mounting process. The harder half is finding out whether they are oriented as intended, with the rotor axle more or less parallel to the surface. It was reassuring to find that the TINK program readily identified numerous local potential energy minima of the hoped-for orientation of all three diastereomeric conformations of the dipolar rotor on the gold surface, but convincing evidence could come only from experiments.

The first condition that the adsorbed molecules need to satisfy is to have the expected surface footprint. In order to find out what size to expect for the desired attachment mode, we have calculated the surface area per molecule for various optimized conformations of the sulfur-containing tentacles using the TINK program [30]. Limiting conformations obtained for the dipolar rotor residing on a gold surface in the desired orientation are shown in Fig. 5.

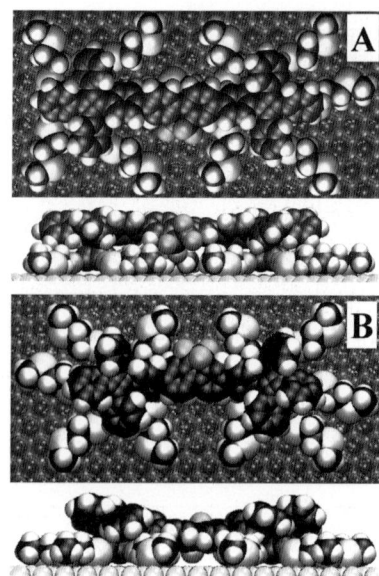

Fig. 5 Limiting tentacle conformations (**A**, one tentacle eclipsed with axle; **B**, tentacles staggered with axle) for gold surface mounted dipolar rotor 2. Top (upper) and side (lower) views

In tentacle conformation A, one of the tentacles is tucked under the rotator, whereas in conformation B, all tentacles are staggered with the axle. Many intermediate conformations of the flexible tentacle chains are obviously possible as well.

In conformation A, one would expect rotation to be blocked and the rotator to be turned with its dipole more or less parallel to the surface. In conformation B, rotation should be possible, and a naive expectation for the favored orientation of the rotator, based on electrostatic interaction with the image charges only, would be with the dipole pointing to the surface or away from it. Clearly, an intrinsic rotational barrier of the biphenyl type is present and will have an influence as well. This issue is addressed below. For either conformation of the rotor's tentacles, its expected surface area per rotor molecule is about 9 nm^2, and for the stand alone, it is about 4 nm^2. Most undesirable orientations of the rotor molecule give smaller surface footprints.

A rough experimental estimate of the area per molecule, both for the dipolar rotor and for the single stand, has been obtained from STM images [30] (Fig. 6), and in both cases it agrees with expectations. The oval shape of the rotor molecule images contrasts with the perfectly circular shape of the single stand molecules. This result in itself does not prove that the rotor molecules are oriented on the surface in the way expected from chemical intuition, with the ten mercury atoms and the twenty tentacles on the gold surface, but it is

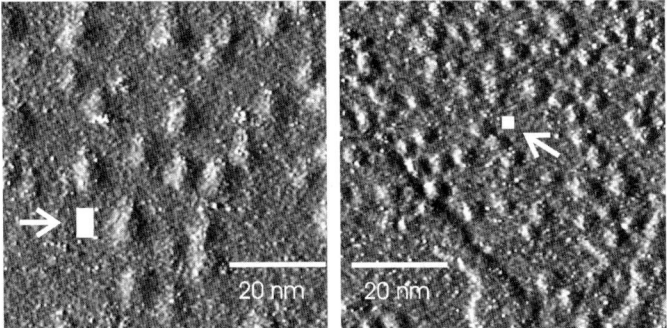

Fig. 6 Derivative mode STM images of rotor molecules **2** (left) and single stands (right) on Au(111). The expected size is shown in white

compatible with this expectation. Interestingly, the rotor molecules deposit in a more or less regular way, often lined up roughly parallel to the Au(111) step edges, and it appears that their adsorption favors a certain arrangement of the surface gold atoms.

An indirect test of the result is possible if one is willing to assume that adsorption on gold will be similar to adsorption on mercury. On mercury, we can use our electrochemical Langmuir trough [21] to obtain reliable compression isotherms, and the surface areas obtained from these, both for the rotor and for the single stand, are 9 and 4.5 nm^2, as expected [30].

3.3
Surface Conformation

The next questions one can ask about the disposition of the molecular rotor on the surface deal with the orientation of the molecular axle, which ought to be parallel to the surface if all is well, and with the orientation of the rotator's dipole relative to the surface normal. It will be reassuring if the dipole has at least some component perpendicular to the surface, which ought to be the electrostatically favored orientation if the rotor resides on the surface in the intended way (Fig. 5).

A method suitable for determining the average orientation of the rotator relative to the gold surface is grazing incidence IR spectroscopy. This takes advantage of the selection rule applicable to molecules adsorbed on the surface of a conductor, according to which only the component of a transition moment that is normal to the surface contributes to absorption intensity. The reason for this is the cancellation of those components of a dipole that are parallel to the surface by the image charges (Fig. 7). Since in an isotropic sample transition moments oriented in any direction are weighted equally, a comparison of peak intensity I^{Au} in the surface spectrum relative to its intensity I^{iso} in the isotropic spectrum depends not only on the total num-

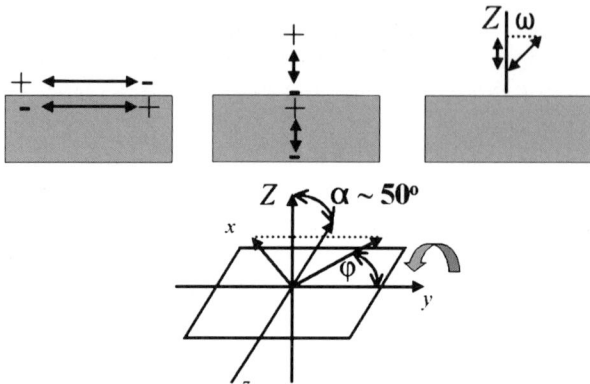

Fig. 7 Origin of the IR transition moment selection rules for an adsorbate on a conductor surface (top). Definition of angles needed for application to rotor 2 (bottom). The rectangle represents the average plane of the rotator, the molecular rotation axle is y, the dipole direction is z, and the surface normal is Z

ber of molecules observed, but also on the average angle of inclination that the IR transition moment makes with the surface normal. In order to use the resulting information about the average alignment of IR transition moments relative to the surface to obtain information about the average alignment of the rotator dipole relative to the surface normal, one needs to know the orientation of the IR transition moments in the molecular frame of the rotator.

The first step therefore needs to be an identification of those IR transitions in the rotor molecule that are localized in the rotator. The choice of C–F bonds in the molecular structure was made with this requirement in mind, since they are known to introduce intense IR bands in the 1000–1300 cm^{-1} region. A comparison of the IR spectra of the dipolar rotator molecule, of the stand without the rotator, and of the 9,9,10,10-tetrafluoro-9,10-dihydrophenanthrene rotator alone, combined with a DFT calculation of IR frequencies and intensities, permitted an identification of seven strong IR bands of the rotator within the rotor molecule. The polarizations of the IR transitions of the rotator were determined by polarization spectroscopy on a sample partially aligned in stretched polyethylene, using procedures worked out previously [48–50]. It was easy to identify three totally symmetric (a) vibrations whose transition dipoles lie in the symmetry axis and coincide with the direction of the permanent molecular dipole, since they all had the same dichroic ratio. The directions of transition moments of the four vibrations that were of symmetry b all lie in a plane perpendicular to the symmetry axis. The angles φ that these transition moments form with the molecular orientation axis (taken to be C2–C7) were determined with only limited accuracy of about ±20°.

The results were compared with those of DFT calculations of the vibrational frequencies and transition moment directions. Those IR transition moment directions in low-symmetry molecules that are not constrained by symmetry to lie in a particular direction are very difficult to calculate accurately, especially if the transitions are relatively weak [51]. In our case, these are the b symmetry vibrations, but good agreement was found, with maximum deviations in the value of φ amounting to about 30°. The situation is probably relatively favorable since the transitions in question are all fairly strong. The transition frequencies and the angles φ are collected in Table 1.

We next assume that the vibrational coupling between the rotator and the stands is weak enough that the transition moment directions in the rotator will be the same in the rotor molecule and in the free rotator itself. Given the surface selection rule, the IR intensity due to a transition moment will be proportional to $\cos^2 \omega$, where ω is the angle between the transition moment direction and the surface normal Z. As shown in Fig. 7, in which the average plane of the rotator is represented schematically by a rectangle, rotation about the axis y of the rotator translates directly into a change of the angle that the rotator symmetry axis z makes with the surface normal, labeled α in the drawing. To obtain the active component of a transition moment of a b symmetry vibration, which lies in the xy plane perpendicular to z, it is necessary to project into x first and then into Z, and this introduces a factor $\sin^2 \varphi$. The ratio I^{Au}/I^{iso} of the intensity of an IR transition in the surface spectrum relative to its intensity in an ordinary isotropic spectrum will be the same for all vibrations of symmetry a, since they are all polarized along z, and is called $I^{Au}(a)/I^{iso}(a)$. However, for the various vibrations of symmetry b, the ratios $I^{Au}(b)/I^{iso}(b)$ will be different unless their angles φ are the same, and they will in general also differ from $I^{Au}(a)/I^{iso}(a)$. If the intensity ratios I^{Au}/I^{iso} can be measured for at least one vibration of symmetry a and one of symmetry b, it will be possible to evaluate

Table 1 Important vibrational modes of 9,9,10,10-tetrafluoro-9,10-dihydrophenanthrene

No.	ν(cm^{-1})	Sym.	φ(exp)	φ(DFT)
1	1259	a	–	–
2	1230	a	–	–
3	1150	b	50	43
4	1091	b	40	70
5	1072	b	70	55
6	1047	b	50	82
7	1006	a	–	–

[a] See text and Fig. 7 for the definition of the angle φ

the average angle α between the dipole direction z and the surface normal Z from the formula $\tan^2 \alpha = [I^{Au}(a)/I^{iso}(a)]/[\sin^2 \varphi I^{Au}(b)/I^{iso}(b)]$. The result will be an average over the orientation of all molecules observed in the sample.

A recent substantial improvement of single-reflection grazing incidence IR sensitivity [52] permitted us to obtain adequate spectra of submonolayers of the rotor molecules adsorbed on gold, as determined by ellipsometry. These are compared with spectra of ordinary isotropic samples in a KBr pellet in Fig. 8. The three vibrations of symmetry a are barely detectable in the surface spectrum. They have identical and small intensity ratios I^{Au}/I^{iso} within the large experimental error, and their average defines $I^{Au}(a)/I^{iso}(a)$. An independent determination of α is possible from each of the four $I^{Au}(b)/I^{iso}(b)$ values measured for the vibrations of symmetry b. The results all agree and yield an average value of $50 \pm 20°$ for α [30]. This result is compatible with the notion that some rotators are free to adopt the electrostatically favorable orientation in which the dipole is normal to the surface as in conformation **A** in Fig. 5 ($\alpha = \sim 0°$), while others are blocked, with a tentacle as in conformation **B**, or with adventitious contaminants, and have their dipole more or less parallel to the surface ($\alpha = \sim 90°$).

The anticipated angle α cannot be evaluated more reliably by molecular mechanics, primarily because the relative energies of the various conformations provided by the UFF force field used by the TINK program are of limited accuracy. The calculations yield relative energies of 0, 3, and 6 kcal/mol for the PPM, PPP, and PMP diastereomers on gold, and for each of them, they favor **B** type conformations (Fig. 5) over **A** type conformations by about 6 kcal/mol [30]. We consider these differences to be of little significance, given the accuracy of the method, and believe that all of these conformations, and many with an intermediate arrangement of the tentacles, are likely

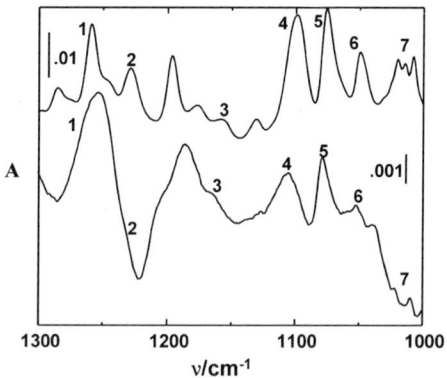

Fig. 8 FTIR spectra of surface-mounted dipolar rotor **2** in a KBr pellet (top) and on Au(111) (bottom). Vibration numbering corresponds to Table 1

to coexist on the gold surface and to interconvert rapidly at room temperature. The calculated barrier to the P-M interconversion of a stand in the surface mounted rotor is 14 kcal/mol, significantly more than in the free molecule, but still small enough for rapid thermal interconversion under ambient conditions.

4
Rotor Performance

The ultimate test of a successful surface-mounted rotor is to determine that it indeed can rotate. We have devised a way to check this on a single-molecule basis and found that at any one time, about one-third of our gold-mounted rotors indeed are free to flip the direction of their dipole in response to an outside electric field and are apparently turning rapidly and randomly at room temperature under the influence of thermal excitation. The other two-thirds of the rotors appear to be blocked. Over a period of tens of minutes, a rotor changes from the blocked state to a state where it is free to rotate, or from the free state to the blocked state.

4.1
Barrier Height Imaging

In order to detect whether any particular molecular rotor on the gold surface can turn in response to an electric field, we have used a modification of STM known as barrier height imaging (BHI). This is a procedure in which the distance z of the tip from the conducting surface is modulated rapidly (in our case, at 5 kHz) by a small amount (in our case, about 0.3 Å). This causes the tunneling current i to be modulated at the same frequency, and both the current and its derivative di/dz are measured (Fig. 9). Since i falls off exponentially with z, so does di/dz. The proportionality constant that relates i and di/dz is equal to $-\beta\phi^{1/2}$, where β is a known constant and ϕ is the combined work function of the surface and the tip. The work function of a metal is the amount of energy needed to remove an electron from its surface to infinity. The physics behind this measurement relies on the fact that the easier it is to remove an electron from the surface, the farther out into vacuum will its wave function extend, and the slower will be the fall off or the tunneling current to the tip as z is increased.

The work function of a surface is strongly affected by adsorbates and especially, by surface dipoles. The strength of the electric field between an STM tip and a surface is extremely high. Since the voltage difference between the tip and the surface is on the order of volts, and the distance between them is on the order of nm, the field strength is on the order of GV/m. This is sufficient to line up a dipole of the size of ours, about 4 D, even at room temperature,

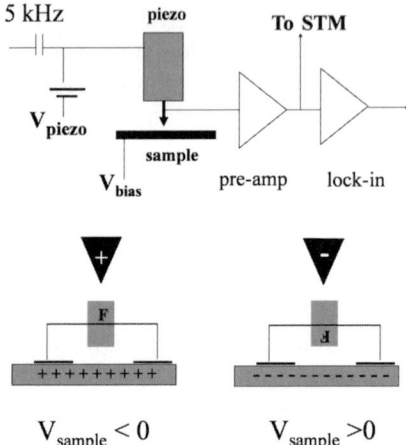

Fig. 9 Barrier height imaging apparatus schematic (*top*) and field-induced dipole flipping (*bottom*)

and keep it oriented over 99% of the time (Fig. 9). Depending on the direction of the field, which we can control, we can therefore hold the negative end of the rotator dipole, where the fluorines are located, close to the surface or far from the surface, as long as it is free to flip. These two situations represent two different conformations of the rotor, and because they differ in the orientation of a surface dipole, they should have detectably different local work functions, ϕ_+ and ϕ_-.

In our differential BHI measurement [30], we raster the tip over the sample as usual, but go over each line twice, once with each field polarity. When the field holds the negative end of the rotator dipole close to the surface, we measure $\phi_+^{1/2}$, and when it is away from the surface, we measure $\phi_-^{1/2}$. In the differential plot, we then show the difference of these two quantities. We can also display the ordinary STM image and pinpoint the location of the rotor molecules on the surface (Fig. 10). When the tip is located over clean gold or over a molecule whose surface dipole does not change its normal component when the electric field direction is reversed, the difference is constant and there is no molecular image in the differential BHI plot. Of course, all molecules are polarizable to some degree and the direction of the small induced dipole will depend on the direction of the outside field, so there will always be a faint image. However, those molecules that respond to the change in field direction by rotating a substantial dipole, such as the 3.7 D dipole of our dipolar rotator, are expected to give strong and clear differential images. This is indeed seen in Fig. 10 for the gold surface covered with dipolar molecular rotors, and not for the control surface covered by the otherwise very similar but non-polar molecular rotors. On the average, about one-third of the dipolar rotors give positive images, and two-thirds are apparently unable

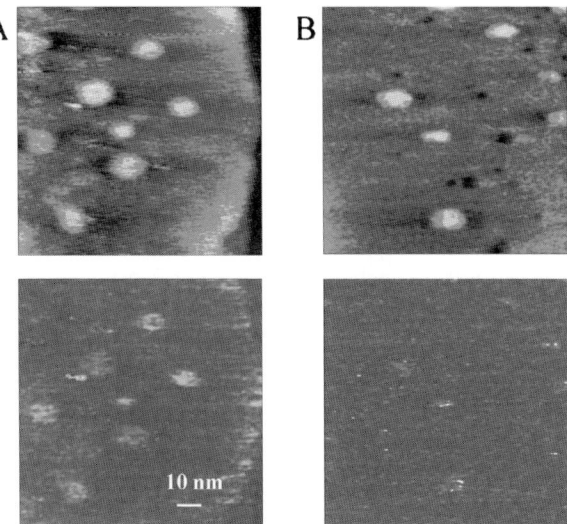

Fig. 10 Constant current STM image (*top*) and differential BHI image (*bottom*) on a Au(111) surface. **A**, dipolar rotors **2**; **B**, non-polar rotors **1**

to turn their dipole when the field reverses direction, since they are blocked for one reason or another, presumably by surface contaminants or by one of their own tentacles.

In the course of tens of minutes, some molecular images become active in the differential scan, while some of those previously active disappear. We attribute this to slow diffusion of the contaminants or the tentacles on the surface. The measurement was done at room temperature under hexadecane, and we assume that in the absence of the locking electric field those rotors that are free to turn oscillate randomly between their favorite orientations. It would be interesting to park the STM tip above a single molecule and perform the differential BHI imaging as a function of temperature. One could then observe the onset of a delayed response to the applied field, and ultimately absence of a response, as the temperature is lowered. This would represent a direct measurement of the rotational barrier. Conceivably, one might even observe transformations between the PPM/MMP, PPP/MMM, and PMP/MPM enantiomer pairs, whose rotational barriers may differ. We now believe these transitions to be too fast to be detected at room temperature, given that it takes about 1/2 s between times at which the tip visits a molecule with opposite field polarities. Such studies of temperature dependence will require an STM instrument capable of operating in an ultrahigh vacuum, and have not been attempted so far.

Even at the present stage, we believe that the results demonstrate that a considerable fraction of our surface-mounted rotors works as intended, in that their rotators are free to flip around their axle fast at room temperature.

4.2
Computation of Demands on the Driving Force

Although we have performed no experiments that would probe the rotor properties in more detail so far, we have asked questions about the behavior to be expected, and used the TINK program to provide tentative answers [30, 53]. The answers appear interesting enough to describe in some detail, even though they depend on approximate computations and must therefore be viewed with a healthy dose of skepticism.

In the experiments described so far, the rotors have been allowed to move as a result of thermal agitation and then were locked in one orientation in an essentially static field for a brief period of time when the STM tip arrived above them, with only occasional brief excursions into the less favorable orientation. It is interesting to ask how a rotor would respond if it were exposed to an alternating field from an STM tip parked permanently above it. Could it be driven into synchronous rotation? Could the rotation be unidirectional? What amplitude of the alternating field would be required to achieve this, as a function of field frequency and temperature? These are the kinds of questions that we have asked earlier about surface-mounted azimuthal rotors [54, 55] and that theoreticians are asking about molecular rotors in general [7].

The first question to address is, what are the effective one-dimensional potentials that the three distinct rotor conformations move in? The true minimum energy paths are not easy to calculate, given the number of atoms in the rotor and the presence of many low-frequency degrees of freedom, and they are less relevant than the actual molecular dynamics calculations. Nevertheless, they may provide some qualitative understanding. An approximation to these potentials is shown in Fig. 11 for all three conformers shown in Fig. 2. Figure 11 also defines the sense of the angle α; when the rotor molecule is viewed along the rotation axis, the plane of the aryl substituent on the stand in front would make an angle of about $-55°$ with the Z axis if the stand has P helicity and about $55°$ if it has M helicity.

The three potentials are quite different and can be understood qualitatively as sums of the electrostatic interaction with image charges in the metal and of barriers imposed by the biphenyl-like attachment at both ends of the rotor. The rotational barriers calculated for these surface-mounted rotors are low, about 3–7 kcal/mol, but somewhat higher than in the free molecule. The electrostatic interactions favor the rotational angle values $\alpha = 0°$ (fluorine substituents as close to the surface as possible) and $\alpha = 180°$. The biphenyl barriers at the two ends favor twist angles of about $35°$ relative to the plane of the aryl substituent on the cyclobutadiene that carries the rotor. These aryl substituents are twisted about $35°$ relative to the cyclobutadiene plane [29] and thus to the gold surface, because they are too crowded around the four-membered ring. Allowing for the small $\sim 15°$ twist between the two aromatic

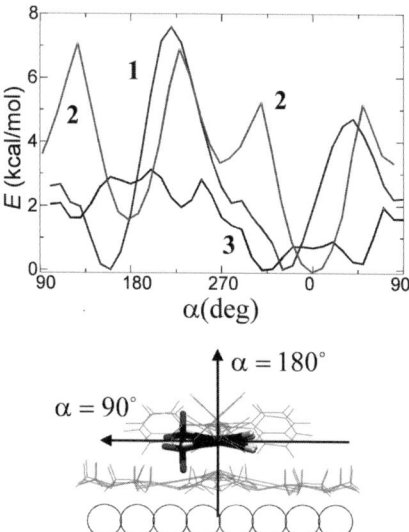

Fig. 11 Top: Calculated rotational barriers of the conformers of the dipolar rotor **2** on Au(111); 1: PPM; 2: PPP; 3: PMP. Bottom: definition of the angle of rotation α (the Au surface is symbolized by a set of circles)

rings of the rotator itself, and depending on the sense of the twist at the rotator attachment point, this factor alone thus favors α values in the vicinity of 0°, −20°, ±90°, 160°, or 180°, if the front stand is P, and the same angles with opposite signs if it is M. The ±90° values are strongly disfavored by the electrostatics. A P (M) stand in the back will have the same effect as an M (P) stand in front. The effects of the stands at the two ends thus work at cross-purposes when the stand handedness is the same on both ends of the rotator. It is then not surprising that the calculated potential curves in Fig. 11 are essentially perfectly symmetric for PPP and nearly symmetric for PMP relative to reflection in the surface normal Z (change of sign of the angle α), with minima at 0° and 180° in the former case and about ±45° and ±135° in the latter.

However, the effects at the two attachment points work hand in hand if the stands are of opposite handedness. Indeed, in the PPM rotor the potential is asymmetric relative to this symmetry operation, and the minima are located at about −20° and 160° when the rotor is viewed with the P stand in front (the signs would change if the M rotor were in front). Thus, for the PPM rotor unidirectional rotation is to be expected even when the driving oscillating electric field is normal to the surface, as it must be when the surface is a conductor. When the field is pulling up the dipole end located close to the surface and pushing down the end that is distant from the surface, it facilitates rotation, and when it does the opposite, it accomplishes little. One would therefore expect a continuous rotation, always in the same sense. The fact that the rotor is chiral is coincidental and if it had a plane of symmetry perpendicular to the rotational axis, it would still rotate unidirectionally as long as the minima were located asymmetrically with respect to the change of sign on the angle α. It appears at first sight that factors that could destroy efficient unidirectional motion might be an elevated temperature, which could impose overwhelming random motion, and a poorly chosen electric field amplitude that could be too small to allow the rotor to overcome the rotational barrier when temperature is too low for thermal jumps. It could also be too large, in which case the effect of the asymmetric potential could become negligible, but this is not a realistic concern in the case of our rotor.

Since the choice of temperature plays a key role in the analysis, we shall separate the discussion of the dynamical results [30, 53] into two regimes, driven and thermal, at low and high temperature, respectively. In the former regime, thermal motion is insufficient to induce rotor hops from one orientation to another and the rotation is driven by the electric field ("driven rotation"). In the latter, thermal motion is sufficient to induce rotor turns, and the electric field merely serves to bias them in one direction ("thermal rotation").

4.2.1
Low Temperature

Figure 12 shows the response of the three rotor conformers kept at 10 K to an electric field of gradually increasing strength oscillating at 90 GHz. At this temperature, the rotors perform essentially no thermal hops. The net number of turns in the expected direction is plotted vertically, with turns in the "wrong" direction subtracted from the turns in the "correct" direction, and

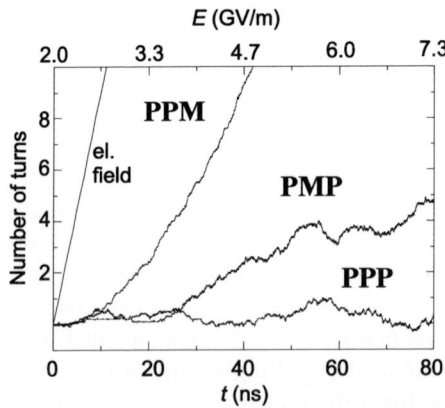

Fig. 12 Calculated unidirectional response of surface-mounted dipolar rotors 2 to 90 GHz electric field. The upper horizontal scale shows how field amplitude is increased in time

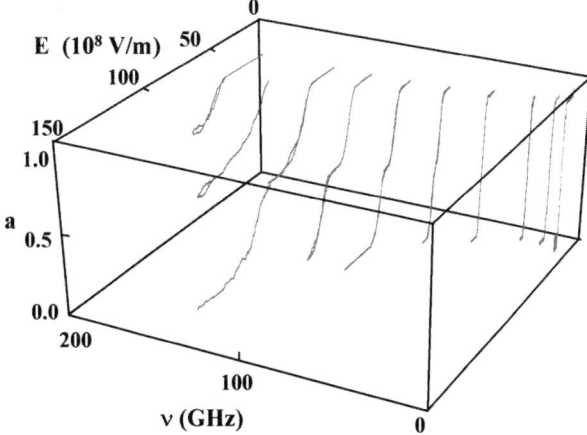

Fig. 13 Calculated phase diagram of surface-mounted dipolar rotor 2 at 10 K

time and field amplitude are both plotted horizontally. It is seen that the PPP (or MMM) rotor shows no directional preference and the PMP (or MPM) rotor a moderate one, while the PPM (or MMP) rotor, whose potential is strongly asymmetric (Fig. 11), exhibits the anticipated strongly unidirectional rotation. The straight line shown counts the cycles of the electric field, and it is apparent that for the stronger fields, the PPM/MMP rotor approaches synchronous behavior. This is the conformer that we have decided to examine in greater detail.

From molecular dynamics runs at many frequencies v and field amplitudes E, we have determined the probability a that the PPM/MMP rotor will skip a cycle of the field (the average lag of the rotor). In Fig. 13, this quantity is plotted against v and E, producing a phase diagram of the rotor at 10 K. When the field is weak, the cycle skip probability a equals one, and the rotor is performing no rotation of any kind, but only librational motion. As the field strength increases and approaches a critical value, the skip probability a drops abruptly, and in strong fields the rotor turns in the correct direction with every cycle of the field. The critical field strength increases with increasing frequency. This is hardly surprising, since at higher frequency the field has less time to force the rotor to flip.

What surprised us is that at frequencies in excess of ~ 60 GHz, a does not drop all the way to 0 immediately, but first dwells at the value of 1/2. Inspection of the trajectories showed that in this regime the motion is half-synchronous, and the rotor regularly skips every other cycle of the field. At frequencies above 150 GHz, there even is an indication of a quarter-synchronous motion. Such behavior is reminiscent of parametric oscillators and has been studied in great detail in non-linear classical mechanics, where it would be called a 2 : 1 resonance. We are not aware of a prior realistic and explicit model for a molecule in which such behavior has been seen.

Can the origin of the half-synchronous response be understood in simple mechanistic terms? In Fig. 14 we show the time-dependent rotational potential for the MMP rotor, which incorporates the effect of an outside electric field with an amplitude of 4.34 GV/m, and display two representative rotor trajectories, at ω = 9 and 90 GHz. At the lower frequency, the rotor essentially follows an adiabatic trajectory. As the well near $\alpha = 20°$ in which it resided at time $t = 0$ is gradually eliminated by the growing electric field and replaced by a well near $\alpha = 200°$ during the first half-cycle of the field, the rotor at first continues to dwell in its original minimum, but then abruptly jumps to the new minimum by performing a half turn. This action is repeated in the second half cycle of the field, and the rotor returns to its original orientation. This sequence of events is repeated during each field cycle and the rotor is synchronous with the field.

The situation is different at 90 GHz. Now, electronic friction felt by the rotor as the image charge moves through the metal makes it lag behind considerably, and it arrives at the first new minimum at $\alpha = 200°$ barely in time before the minimum disappears. The lag increases in the second half-cycle of the field, and although the rotor completes a full turn, it misses the $\alpha = 20°$ minimum altogether. In the next half-cycle, it then has no opportunity to reach the minimum at $\alpha = 200°$ and during the next half-cycle it returns meekly to the original well at $\alpha = 20°$, only to repeat the experience during the next two field cycles. It misses every other cycle of the field and rotates at half frequency.

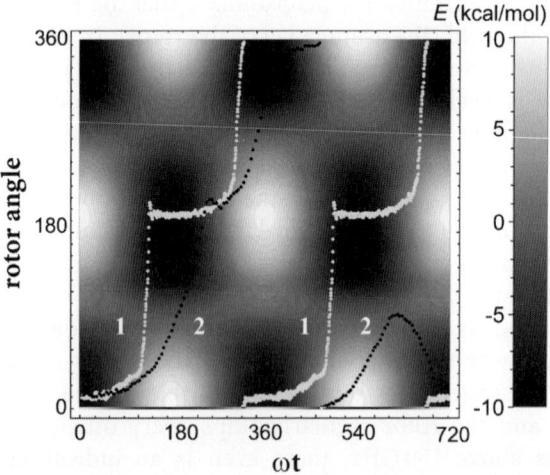

Fig. 14 Representative trajectories calculated for synchronous (9 GHz, curve 1) and half-synchronous (90 GHz, curve 2) motion of gold surface mounted dipolar rotor **2** in an oscillating electric field of 4.34 GV/m amplitude. The shade of grey indicates potential energy (see scale on the right)

4.2.2
Elevated Temperatures

What happens at higher temperatures, when thermal energy permits the rotor to hop between its favored orientations? In the absence of electric field, these hops will occur with equal likelihood in either direction, and there will be no unidirectional rotation ($a = 0$). As the electric field increases, it will bias the direction of the hops and the rotation will become partly unidirectional. This situation has been the subject of much attention in the literature [7] and is usually referred to as the "Brownian motor".

Figure 15 shows the plot of a against E at 90 GHz and a series of temperatures. It is seen that at high field amplitudes, sufficient to drive the rotor synchronously at 10 K, random thermal motion increases the skip probability a and hurts the performance of the unidirectional rotor. This is easy to understand, as random thermal hops can occur with some probability in either direction, whereas the driven motion is strictly unidirectional. This is best understood with the aid of Fig. 16, which shows the time-dependent rotational potential and an opportunity for mischief upon thermal excitation.

At weak fields, insufficient to drive the rotor synchronously at 10 K, the situation is entirely different. Now, the rotor does not perform at all at temperatures that do not permit thermal hops, and $a = 1$. At elevated temperatures, at least some hops become possible, and more of them are in the correct direction than the wrong direction. Hence, $a < 1$ and thermal motion improves the unidirectional performance of the rotor. This is again best understood with the aid of Fig. 16, which shows how the relatively weak per-

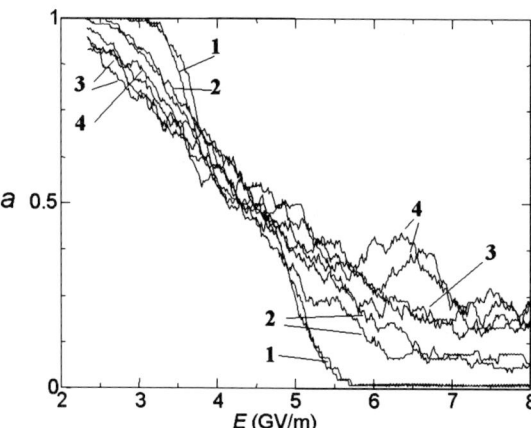

Fig. 15 Calculated temperature dependence of the unidirectional performance of surface-mounted dipolar rotor 2 at 90 GHz, as measured by the skip probability a (1: 10 K; 2: 100 K; 3: 200 K; 4: 300 K)

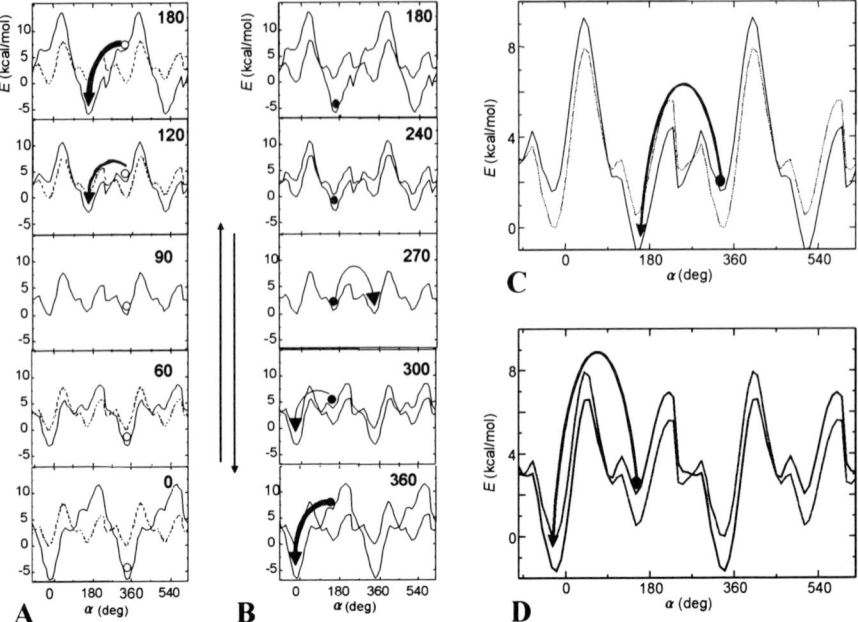

Fig. 16 Calculated time-dependent rotation potentials dictating the driven (**A**, **B**, strong field) and thermal (**C**, **D**, weak field) response of the PPM conformer of the surface-mounted dipolar rotor **2**. The dashed line indicates the potential in the absence of the field. In **A** and **B**, vertical arrows indicate the direction of the change of the electric field and the numbers give its phase (zero corresponds to the strongest field along the surface normal). The field changes direction between **C** and **D**. The orientation of the rotor is indicated by a black dot and its motion by curved arrows

turbation by the electric field modulates the rotational barriers and induces unidirectionality in thermal motion over them.

In summary, the performance of a driven unidirectional rotor is degraded by random thermal motion, whereas the performance of a thermal unidirectional rotor relies on it.

5
Summary and Outlook

In this chapter, we have attempted to outline the steps involved in the fabrication and verification of a surface-mounted altitudinal rotor and in the examination of its performance in electric fields. We have also indicated some of the directions in which future work with electric-field driven rotors of this kind may proceed.

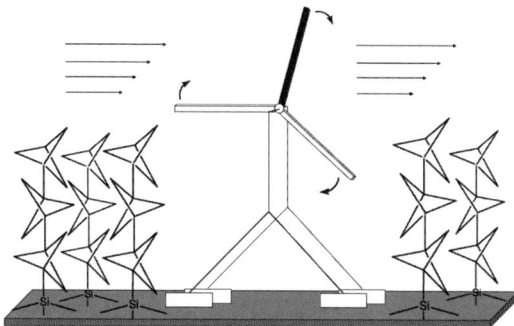

Fig. 17 A possible design of a fluid flow driven altitudinal rotor half embedded in a staffane-based self-assembled monolayer

There are other ways in which surface-mounted altitudinal rotors could be driven, and we shall conclude by briefly mentioning one. This is the use of the flow of gas or perhaps even liquid in a direction parallel to the surface and perpendicular to the rotor axis. A schematic representation of this possibility is shown in Fig. 17. We do not know what rotor design it would take to make it function, and are presently performing molecular dynamics simulations to answer this question.

Acknowledgements Our work on molecular rotors has been supported by the USARO (DAAD19-01-1-0521) and the NSF (CHE-0446688).

References

1. Kinosita K Jr, Yasuda R, Noji H, Ishiwata S, Yoshida M (1988) Cell 93:21
2. Schliwa M (ed) (2003) Molecular Motors. Wiley, Weinheim
3. Feynman RP (1960) Saturday Rev 43, April 2, p 45
4. Feynman R, Leighton RB, Sands M (1963) The Feynman Lectures on Physics, vol. 1. Addison-Wesley Longman, Reading, Mass, Vol 1, 46-1
5. Christie GH, Kenner J (1922) J Chem Soc 121:614
6. Kemp JD, Pitzer KS (1936) J Chem Phys 4:749
7. For a review, see Kottas GS, Clarke LI, Horinek D, Michl J (2005) Chem Rev 105:1281
8. Iwamura H, Mislow K (1988) Acc Chem Res 21:175
9. Feringa BL (2001) Acc Chem Res 34:504
10. Leigh DA, Wong JKY, Dehez F, Zerbetto F (2003) Nature 424:174
11. Kelly TR (2001) Acc Chem Res 34:514
12. Arnold BR, Balaji V, Downing JW, Radziszewski JG, Fisher JJ, Michl J (1991) J Am Chem Soc 113:2910
13. Kaszynski P, Friedli AC, Michl J (1992) J Am Chem Soc 114:601
14. Tinkertoy is a trademark of Playskool, Inc., Pawtucket, RI 02862, and designates a children's toy construction set consisting of straight wooden sticks and other simple elements insertable into spool-like connectors

15. Kaszynski P, Michl J (1988) J Am Chem Soc 110:5225
16. Michl J, Kaszynski P, Friedli AC, Murthy GS, Yang H-C, Robinson RE, McMurdie ND, Kim T (1989) In: de Meijere A, Blechert S (eds) Strain and Its Implications in Organic Chemistry. NATO ASI Series, vol 273. Kluwer Academic Publishers, Dordrecht, The Netherlands, p 463
17. Harrison RM, Magnera TF, Vacek J, Michl J (1997) In: Michl J (ed) Modular Chemistry. Kluwer, Dordrecht, The Netherlands, p 1
18. Vacek J, Michl J (1997) New J Chem 21:1259
19. Prokop A, Vacek J, Michl J (unpublished results)
20. Clarke LI, Horinek D, Kottas GS, Varaksa N, Magnera TF, Hinderer TP, Horansky RD, Michl J, Price JC (2002) Nanotechnology 13:533
21. Varaksa N, Pospíšil L, Magnera TF, Michl J (2002) Proc Nat Acad Sci USA 99:5012
22. Magnera TF, Michl J (2002) Proc Nat Acad Sci USA 99:4788
23. Aviram A, Ratner MA (eds) (1998) Molecular Electronics: Science and Technology. New York Academy of Sciences, New York, NY
24. Troisi A, Ratner MA (2002) J Am Chem Soc 124:14528
25. Troisi A, Ratner MA (2004) Nano Letters 4:591
26. Rappe AK, Casewit CJ, Colwell KS, Goddard III WA, Skiff WM (1992) J Am Chem Soc 1/4:10024
27. Rappe AK, Goddard III WA (1991) J Phys Chem 95:3358
28. Schwab PFH, Levin MD, Michl J (1999) Chem Rev 99:1863
29. Harrison RM, Brotin T, Noll BC, Michl J (1997) Organometallics 16:3401
30. Zheng X, Mulcahy ME, Horinek D, Galeotti F, Magnera TF, Michl J (2004) J Am Chem Soc 126:4540
31. Schöberl U, Magnera TF, Harrison R, Fleischer F, Pflug JL, Schwab PFH, Meng X, Lipiak D, Noll BC, Allured VS, Rudalevige T, Lee S, Michl J (1997) J Am Chem Soc 119:3907
32. Mulcahy ME, Magnera TF, Michl J (unpublished results)
33. Zheng X, Wang B, Michl J (unpublished results)
34. Zhong C, Brush R, Anderegg J, Porter M (1999) Langmuir 15:518
35. Noh J, Kato H, Kawai M, Hara M (2002) J Phys Chem B 106:13268
36. Ishida T, Abe K, Yase K, Tamada K (2000) Langmuir 16:1703
37. Baxter PNW (1996) In: Lehn JM, Chair E, Atwood JL, Davis JED, MacNichol DD, Vögtle F (eds) Comprehensive Supramolecular Chemistry, vol 9, Chapt 5. Pergamon Press, Oxford, p 165
38. Fujita M (1998) Chem Soc Rev 27:417
39. Balzani V, Gomez-Lopez M, Stoddart JF (1998) Acc Chem Res 31:405
40. Caulder DL, Raymond KN (1999) Acc Chem Res 32:975
41. Leininger S, Olenyuk B, Stang PJ (2000) Chem Rev 100:853
42. Lehn J-M, Ball P (2000) New Chem 300
43. Cotton FA, Lin C, Murillo CA (2001) Acc Chem Res 34:759
44. Schwab PFH, Noll BC, Michl J (2002) J Org Chem 67:5476
45. Caskey DC, Shoemaker RK, Michl J (2004) Org Lett 6:2093
46. Caskey DC, Michl J (2005) J Org Chem 70:5442
47. Caskey DC, Wang B, Zheng X, Michl J Collect Czech Chem Commun (in press)
48. Michl J, Thulstrup EW (1987) Acc Chem Res 20:192
49. Thulstrup EW, Michl J (1988) Spectrochim Acta 44A:767
50. Michl J, Thulstrup EW (1995) Spectroscopy with Polarized Light. Solute Alignment by Photoselection, in Liquid Crystals, Polymers, and Membranes, revised soft-cover edition. Wiley, Weinheim, p 592

51. Radziszewski JG, Downing JW, Gudipati MS, Balaji V, Thulstrup EW, Michl J (1996) J Am Chem Soc 118:10275
52. Mulcahy ME, Berets SL, Milosevic M, Michl J (2004) J Phys Chem B 108:1519
53. Horinek D, Michl J (2005) Natl Acad Sci 102:14175
54. Vacek J, Michl J (2001) Proc Natl Acad Sci USA 98:5481
55. Horinek D, Michl J (2003) J Am Chem Soc 125:11900

Towards a Rational Design of Molecular Switches and Sensors from their Basic Building Blocks

Nicolle N. P. Moonen · Amar H. Flood · Juan M. Fernández · J. Fraser Stoddart (✉)

California NanoSystems Institute and Department of Chemistry and Biochemistry, University of California, Los Angeles, 405 Hilgard Avenue, Los Angeles, CA 90095-1569, USA
stoddart@chem.ucla.edu

1	Introduction	101
1.1	Timeline for the Development of Rotaxane-Based Switches	101
1.2	Principles of Thermodynamics and Kinetics of Host-Guest Complexes and Mechanically Interlocked Molecules	107
2	Chemical Tuning of the Association Constants	109
2.1	Donor Strength	110
2.2	[C–H···π] and [C–H···O] Interactions	110
2.3	π-Surfaces	111
2.4	Solvent Effects	112
3	Shuttling in Degenerate [2]Rotaxanes—Molecular Shuttles	112
3.1	Influence of K_a on Shuttling Rate	112
3.2	Influence of the Spacers	113
4	Bistable [2]Rotaxanes—Molecular Switches	114
4.1	Equilibrium Population Ratios	115
4.1.1	Solvent Effects	116
4.1.2	Counterion Effects	118
4.1.3	Thermodynamic Parameters	118
4.1.4	Correlation between Thermodynamics and Physical Performance	120
4.2	Chemical Tuning of Switching Speeds ($\Delta G^{\ddagger}$)	122
4.3	Switching in Bistable Rotaxanes	124
5	Environmental Effects	125
6	Conclusions	128
7	Epilogue	128
	References	129

Abstract A fundamental understanding of the thermodynamics and kinetics of mechanically interlocked molecules, such as [2]rotaxanes, will contribute to a more rational design of new molecular machines. This Chapter describes the influence of chemical modifications and the role of the physical environment on the ground state thermodynamics and the shuttling and switching kinetics of several tetrathiafulvalene- and

1,5-dioxynaphthalene-containing [2]rotaxanes. A comparison between the properties of these bistable rotaxanes and model host-guest complexes of the corresponding π-electron donating recognition units with the π-electron accepting cyclophane, cyclobis(paraquat-p-phenylene), has been made, resulting in useful guidelines for the design of new bistable rotaxanes with specific, desirable physical performances.

Keywords Kinetics · Molecular shuttles · Rotaxanes · Thermodynamics · Translational isomerism

Abbreviations

BPTTF	bis(pyrrolo)tetrathiafulvalene
CT	charge-transfer
CV	cyclic voltammetry
E_{ox}	oxidation potential
Et	ethyl
ΔG	free energy difference
ΔG_a	free energy of association
$\Delta G^\ddagger$	free energy of activation
ΔH	enthalpy difference
$\Delta H^\ddagger$	enthalpy of activation
ΔS	entropy difference
$\Delta S^\ddagger$	entropy of activation
DEG	diethylene glycol
DNP	1,5-dioxynaphthalene
g	gram(s)
GSCC	ground-state co-conformation
h	Planck constant (6.262×10^{-34} J s)
I	current
ITC	isothermal titration calorimetry
J	Joule
k	rate constant
K	Kelvin
K	constant
K_a	association constant
k_B	Boltzmann constant (1.381×10^{-23} m^2 kg s^{-2} K^{-1})
K_{eq}	equilibrium constant (N_{GSCC}/N_{MSCC})
M	molarity
Me	methyl
min.	minute(s)
mol	mole(s)
MPTTF	mono(pyrrolo)tetrathiafulvalene
MSCC	metastable state co-conformation
MSTJ	molecular-switch tunnel junction
N	population
NEMS	nanoelectromechanical systems
NMR	nuclear magnetic resonance
PES	potential energy surface
R	universal gas constant (8.3144 J mol^{-1} K^{-1})
r.t.	room temperature

s	second(s)
SAM	self-assembled monolayer
T	temperature
TRISPHAT⁻	tris(tetrachlorobenzenediolato)phosphate(V)
TTF	tetrathiafulvalene
UV/Vis	ultraviolet/visible
VT	variable temperature
VTT CV	variable time and temperature cyclic voltammetry

1
Introduction

Molecular machines have found increasing applicability in nanoscale devices. Examples include molecular electronics [1], nanoelectromechanical systems (NEMS) [2–12], nanovalves [13], and electron relays [14, 15].

The design of a molecular machine is based on an understanding of the molecular machine's final usage, e.g., (i) bistability (0,1) for molecular electronics [1]; (ii) a palindromically displayed pair of mobile rings [2, 3] to mimic the extension/contraction of a muscle [9–12]; (iii) modification of end groups: self-organization as Langmuir–Blodgett films using amphiphilic groups [1, 16, 17], or disulfides for the formation of self-assembled monolayers (SAMs), and covalent attachment on SiO_2 surfaces using hydroxyl groups [13]. However, in all cases an essential feature of the molecular design has been *efficient switching* and *bistability*—the basis [18, 19] for which has emerged over the past 25 years. Underpinning each design has been the necessity for efficient syntheses of mechanically interlocked molecules. These syntheses rely heavily on supramolecular assistance to covalent bond formation [20] and hence on the practical use of thermodynamic factors [21] controlling self-assembly and templation [22, 23]. These same thermodynamic factors live on in the mechanically interlocked molecules and emerge as key properties that control their switching behavior. Similar approaches to the design and optimization have been investigated for rotary chiroptical switches [24].

This review will cover the development of switches based on mechanically interlocked molecules and their thermodynamic [25–27] and kinetic aspects [28–30]. These aspects will be discussed with reference to (1) degenerate (molecular shuttles) and bistable [2]rotaxanes; (2) the switching of bistable [2]rotaxanes; and (3) the influence of the environment on the behavior of these systems.

1.1
Timeline for the Development of Rotaxane-Based Switches

Our group's contribution to the emerging area of molecular machines begins [31] with the synthesis of the π-electron-deficient cyclobis(paraquat-p-phenylene) cyclophane 1^{4+} (Fig. 1). There are two key properties of 1^{4+} which allowed the development of a wide range [32–40] of new structures: (1) the cyclophane interacts with π-electron rich guests by π–π stacking and charge-transfer (CT) interactions [41–45], and with appropriate derivatives of such guests also by [C–H···O] interactions [46–54]; (2) the presence of a rigid cavity helps the cyclophane to trap the guests, affording inclusion complexes. Thus, both electronic and steric effects have made this particular cyclophane very attractive for applications related to molecular machinery. In its infancy, the cyclophane was used in the context of host-guest chemistry [55]. Its full potential as a vital part of molecular machinery was reached, however, when it was used to construct [2]catenanes [56] and [2]rotaxanes [18, 57]. When rendered bistable, [2]rotaxanes are a central component in numerous molecular switches, harnessed to date in molecular electronics [1]. It is their readiness, however, to do real mechanical work in NEMS that has been exploited in a few choice cases [2–6] of late.

In 1988, the host-guest chemistry of the tetracationic cyclophane 1^{4+} was established [55]. These early studies involved the complexation of 1,4-dimethoxybenzene (2, Fig. 2) by 1^{4+}. In the 1 : 1 complex, CT and π–π stacking interactions between the π-electron donor and acceptor contribute to the noncovalent bonding that drives the formation (Scheme 1) of the inclusion complex $[2 \subset 1]^{4+}$. The association constant (K_a) between the tetracationic cyclophane 1^{4+} and the guest 2 in MeCN at 298 K was found [58] to be 18 M^{-1}. Furthermore, the crystal structure of the 1 : 1 complex revealed [55] that the neutral hydroquinone guest is inserted centrosymmetrically through the middle of the tetracationic cyclophane's cavity. This discovery of an inclusion complex led to an exploration of the types of guests that are recognized by 1^{4+}. It was found that this host has an excellent receptor for a wide range of guests (2–10, Fig. 2) that contain π-electron-rich ring systems. For example, tetrathiafulvalene (TTF, 3), a well-known [59] π-electron donor, was found [60] to form a 1 : 1 inclusion complex with 1^{4+} in both solution and in the solid state. TTF undergoes two consecutive one-electron oxidation processes, a key property that anticipates its use in the constructing of molecular

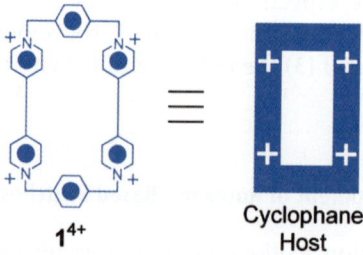

Fig. 1 Cyclobis(paraquat-p-phenylene) cyclophane 1^{4+}

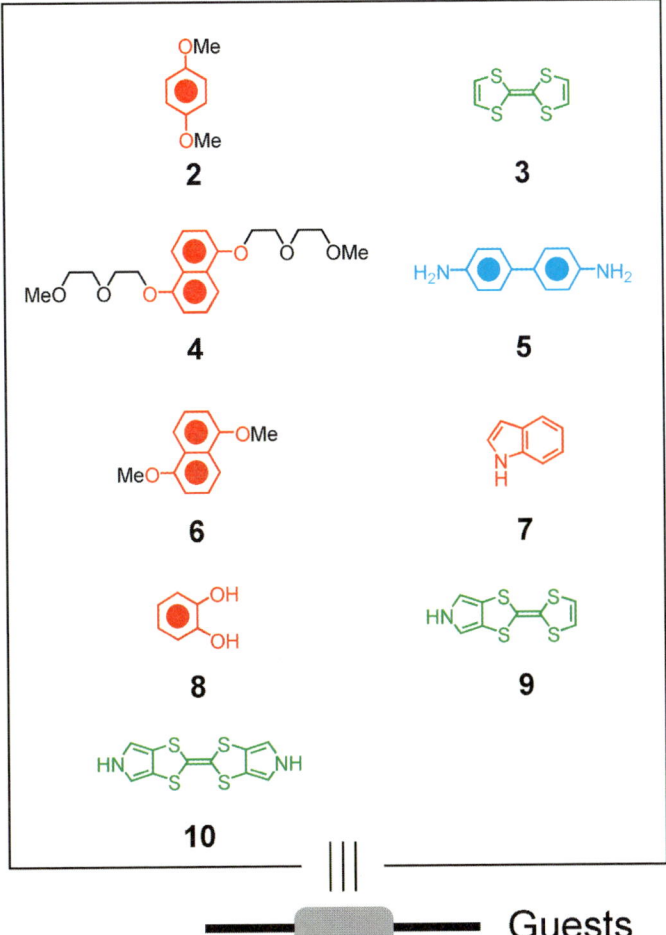

Fig. 2 π-Electron-donating guests 2–10

machines with electrochemically controllable internal movements of its components [16, 61–65]. The K_a for the association between 1^{4+} and TTF (**3**) is 6900 M^{-1} (MeCN) [66] which makes it an excellent guest for the tetracationic cyclophane. However, the binding strength between guests and 1^{4+} is not solely determined by the strength of the π-donor. Subtle structural differences and substituents attached to the donor's core, in particular diethylene glycol (DEG) chains, were also found (*vide infra*) to play a surprisingly dramatic role in determining the binding strengths of these 1 : 1 complexes.

The ability of 1^{4+} to form strong complexes with π-electron-rich substrates was recognized as providing an opportunity to use appropriate donors as templates to direct the formation of the cyclophane itself. Subsequently, the template-directed synthesis [22, 23] proved to be a very efficient strategy in optimizing the yield of 1^{4+}. For instance, in the presence of a template, such as the 1,5-dioxynaphthalene (DNP)-based polyether **4**, a remarkably high yield of 81% of 1^{4+} was obtained [67].

The ability of the tetracationic cyclophane 1^{4+} to form inclusion complexes provides a unique opportunity to construct large and ordered molecular assemblies such as rotaxanes. Simple [2]rotaxanes can be created from a dumbbell compound, containing a π-electron-donating unit, which interlocks the π-electron-deficient cyclophane. These rotaxanes can be self-assembled using two different approaches: clipping and threading-followed-by-stoppering (Scheme 1). Clipping refers to the formation of the tetracationic cyclophane around the dumbbell component utilizing a template-directed protocol [67]. An alternative approach relies on threading of an unstoppered dumbbell through the cavity of the tetracationic cyclophane to form a [2]pseudorotaxane. The resulting [2]rotaxane is formed by attaching covalently bulky stoppers onto the ends of the rod (Stoppering, Scheme 1b). In both cases, the template effect of the π-electron-rich guest dictates the outcome of the reactions to give the desired [2]rotaxanes.

A two-station, degenerate [2]rotaxane can be prepared by adding a second and identical π-electron-rich donor into the dumbbell component. The presence of a second station enables shuttling of the tetracationic cyclophane to

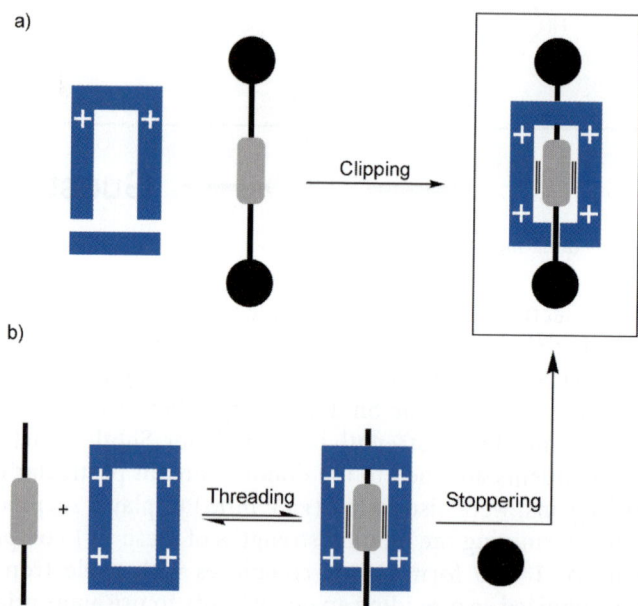

Scheme 1 **a** Two strategies for preparing a rotaxane: **a** clipping and **b** threading-followed-by-stoppering

occur along the rod section of the dumbbell. The first molecular shuttle **11**$^{4+}$, which relies upon this design, was prepared [68] in 32% yield by clipping (Scheme 2) the two components (**12**$^{2+}$ and **13**) of the cyclophane around the two-station dumbbell **14**. At r.t. in CD$_3$COCD$_3$, the tetracationic cyclophane was found to move back and forth like a shuttle between the two identical hydroquinone rings on the rod with a frequency of 500 times per second, a rate which corresponds to an energy barrier of 13.0 kcal mol^{-1}. This shuttle has served as the prototype for the construction of more intricate molecular machines.

The shuttling gives rise to two different degenerate co-conformations [69] that are in dynamic equilibrium. The introduction of two donors with differing affinities for the cyclophane offers the opportunity to differentiate between the two co-conformations, i.e., one of the two possible translational isomers becomes preferred at equilibrium. Furthermore, by selec-

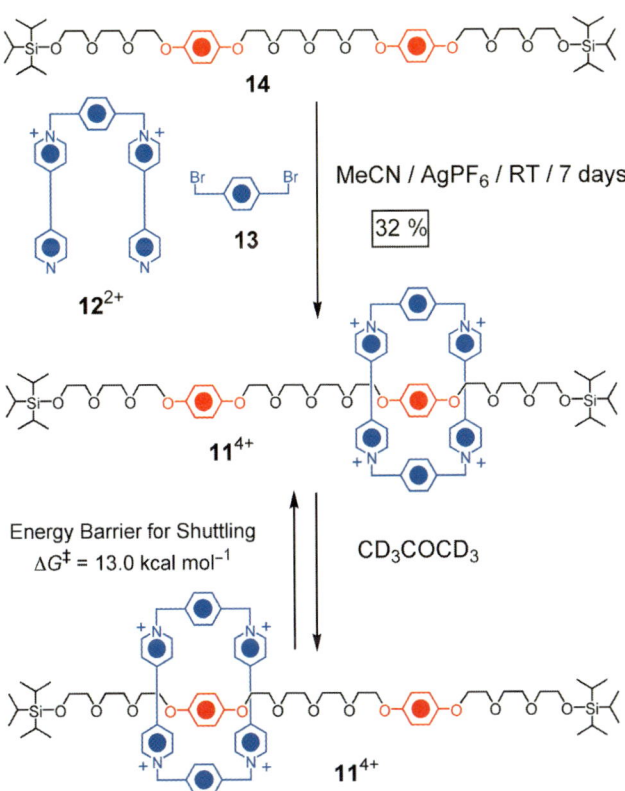

Scheme 2 Templated synthesis of the [2]rotaxane **11**$^{4+}$ and the shuttling process of the tetracationic cyclophane between the two degenerate recognition sites

tively switching the binding between the more preferred donor unit and the electron-accepting cyclophane off and on, the opportunity arises to control the movement of the cyclophane.

The first switchable [2]rotaxane that we reported (**15**$^{4+}$, Scheme 3) incorporated [69] a benzidine and a 4,4′-biphenol unit as the two π-electron-rich donor units. The [2]rotaxane **15**$^{4+}$ exits as two translational isomers in a ratio of 84 : 16 in CD$_3$CN solution at −4 °C, with the tetracationic cyclophane located preferentially on the more π-electron-rich benzidine ring system. When an excess of deuterated trifluoroacetic acid is added to the solution, only one translational isomer [**15**·2D]$^{6+}$, in which the 4,4′-biphenol residue is encircled by the cyclophane, is observed. This dramatic change in structure is brought about by protonation of the basic nitrogen atoms associated with the benzidine ring system. Addition of deuterated pyridine returns the solution to neutrality, and reinstates the previous distribution of translational isomers in **15**$^{4+}$. Furthermore, this switchable molecular shuttle can also undergo reversible electrochemical switching. Mono-oxidation of the more

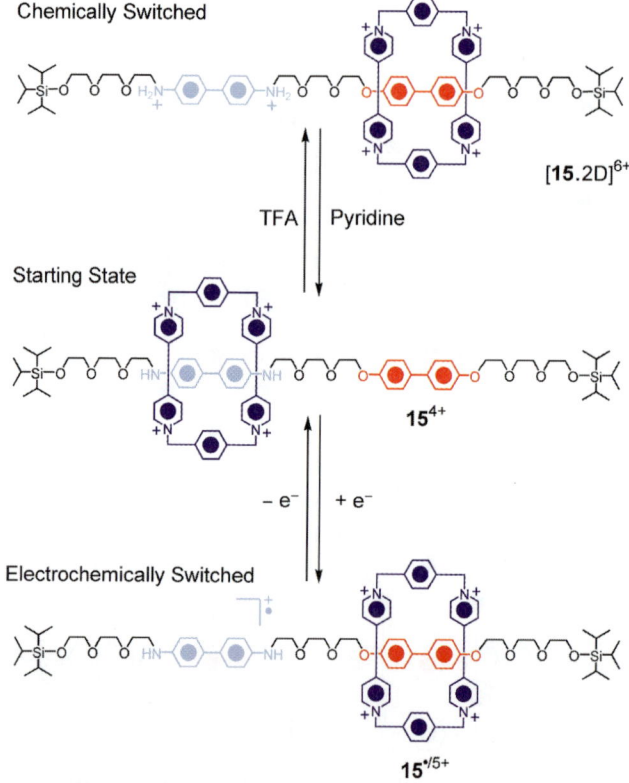

Scheme 3 Chemical and electrochemical switching of the [2]rotaxane **15**$^{4+}$

π-electron-rich benzidine residue switches the structure to one in which the radical form **15**$^{\cdot/5+}$ in which the 4,4′-biphenol unit is selectively encircled by the tetracationic cyclophane. Both the chemical and electrochemical switching processes are completely reversible. This [2]rotaxane demonstrates the reversible on-off switching of one of the molecular recognition units, concomitant with the movement of the cyclophane ring from one station to the other.

1.2
Principles of Thermodynamics and Kinetics of Host-Guest Complexes and Mechanically Interlocked Molecules

The thermodynamics and kinetics of host-guest complexes and mechanically interlocked molecules will be discussed with reference to potential energy surface (PES) diagrams (Figs. 3 and 4). In the case of the host-guest complexes, these diagrams describe the interaction of a π-electron-donating guest, which may be located along a rod component, with a π-electron poor cyclophane host, resulting in a stable complex, that is reflected (Fig. 3) by an energy minimum in the PES. The association constant K_a for the complex is defined as the ratio between the concentration of the complex, divided by the concentration of the free components (Eq. 1). The K_a value is exponentially dependent on the difference in free energy (ΔG_a) between the separate

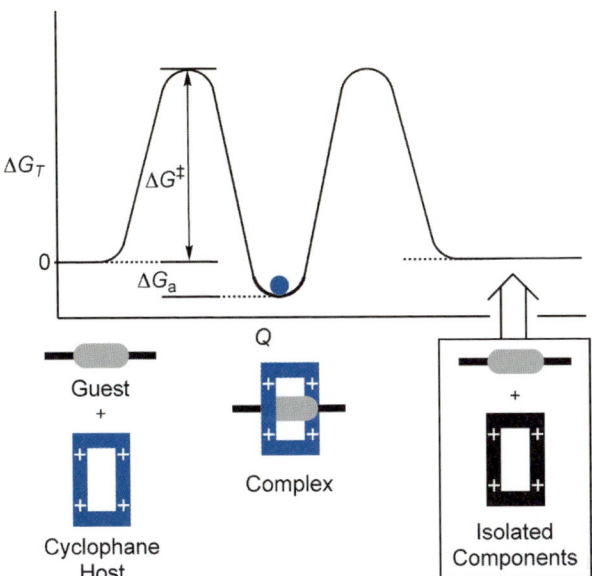

Fig. 3 Potential energy surface diagram for the complexation of a guest with a cyclophane host. The zero of energy is defined as the energy of the isolated components

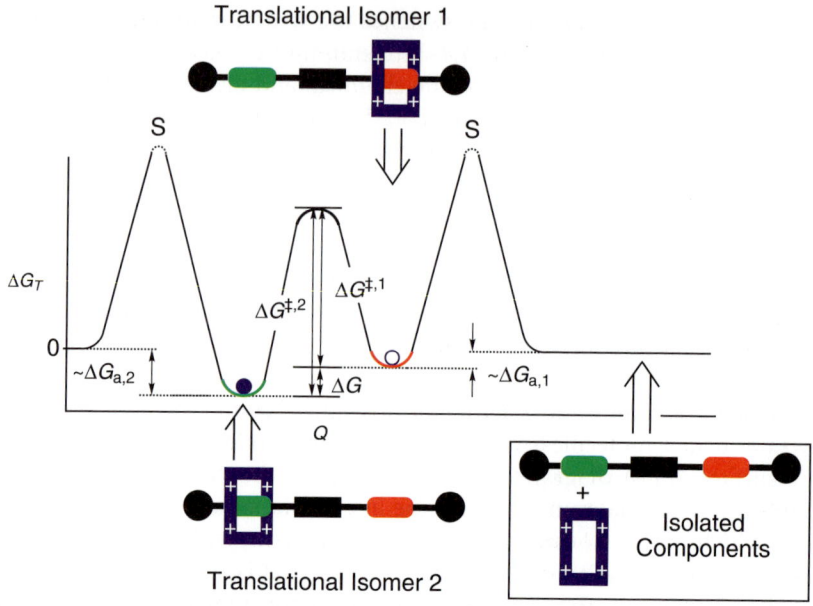

Fig. 4 Potential energy surface diagram of a bistable [2]rotaxane defining the free energy difference ΔG between the two translational isomers (1 and 2) and the barriers $\Delta G_i^\ddagger$ to convert between the two of them. The energy of the bottom of each well ($\Delta G_{a,i}$) that is associated with the translational isomers is defined with respect to a common reference point of the isolated components and is related to the free energy of complexation of the model guests with the cyclophane. The energy barriers of the stoppers (S) are much larger than for the barrier to isomerization

components and the complex, divided by the temperature, T.

$$K_a = \frac{[\text{Complex}]}{[\text{Cyclophane Host}][\text{Guest}]} = e^{-\frac{\Delta G_a}{RT}} \quad (1)$$

$$\Delta G = \Delta H - T\Delta S \quad \text{Gibbs–Helmholtz} \quad (2)$$

$$\frac{\Delta G}{T} = -R \ln K = \frac{\Delta H}{T} - \Delta S \quad \text{Van 't Hoff} \quad (3)$$

R = Universal Gas Constant (8.3144 J mol^{-1} K^{-1}).

The free energy barrier to complexation $\Delta G^\ddagger$ is related to the complexation rate constant k (Eq. 4):

$$k = e^{-\frac{\Delta G^\ddagger}{RT}} \quad (4)$$

$$\frac{\Delta G^\ddagger}{T} = -R \ln\left(\frac{hk}{k_B T}\right) = \frac{\Delta H^\ddagger}{T} - \Delta S^\ddagger \quad \text{Eyring} \quad (5)$$

h = Planck constant (6.262 × 10^{-34} J s)
k_B = Boltzmann constant (1.381 × 10^{-23} m^2 kg s^{-2} K^{-1}).

These two free energy parameters—ΔG and $\Delta G^{\ddagger}$—will play a key role in the discussion of the thermodynamic and kinetic properties of the mechanically interlocked molecules in this review. For interlocked molecules, such as [2]rotaxanes, the cyclophane is mechanically constrained along the rod section of a dumbbell bearing large bulky stoppers, leading to significant free energy barriers. In such cases, the appropriate PES, defining the space between each stopper, reflects the interaction of the cyclophane with the recognition sites on the rod section of the dumbbell (Fig. 4). With this definition, the location of the bottom of the well is related to the separated components, i.e., to a first order approximation by ΔG_a. For bistable rotaxanes, this model results in a double-welled surface with an intervening barrier, $\Delta G^{\ddagger}$. One translational isomer is assigned to each well and the more favored isomer is called the ground-state co-conformation (GSCC) and the less favored one, the metastable state co-conformation (MSCC). The two recognition sites and co-conformations lead to bistability. The free energy difference between the wells, ΔG, is a measure of the difference in the affinity of the cyclophane for each unit and is related to host-guest complexation, as the difference between the free energies of association, $\Delta G \sim \Delta(\Delta G_a)$, of each model guest with the cyclophane. The equilibrium constant, K_{eq}, for the bistable rotaxane is defined as the ratio between the population of one translational isomer over the other, N_{GSCC}/N_{MSCC} (Eq. 6).

$$K_{eq} = \frac{N_{GSCC}}{N_{MSCC}} \tag{6}$$

This review will focus primarily on [2]rotaxanes which consist of the π-electron accepting cyclophane 1^{4+} (in blue), behaving in the manner of a host along the single-coordinate of the dumbbell that bears two π-electron donors based on TTF derivatives (in green) and/or DNP units (in red) as guests (Fig. 4) separated by a spacer (black).

For these bistable [2]rotaxanes, $\Delta G^{\ddagger}$ is in general > 13 kcal mol^{-1} and $\Delta G \leq 4$ kcal mol^{-1}. More detailed insight into the relationship between the chemical structure and the physical performance of these systems is provided by the analysis of the enthalpy (ΔH) and entropy (ΔS) of equilibrium and activation from the Gibbs–Helmholtz equation (Eq. 2). The van 't Hoff equation (Eq. 3) describes the influence of these quantities on the association (K_a) and equilibrium (K_{eq}) constant. The Eyring equation (Eq. 5) is used to determine the thermodynamic parameters of activation.

2
Chemical Tuning of the Association Constants

The thermodynamic parameters and therefore the physical performance of the bistable rotaxanes are related to the binding constant K_a (see Eq. 3) of the appropriate host-guest complexes. Numerous investigations have been performed on the factors determining the complexation strength of supramolecular complexes. Here, we will summarize the most important aspects for the complexation of the cyclophane host with electron-donating guests.

2.1
Donor Strength

The donor strength of the guest molecule is an important factor in its complexation with the electron poor cyclophane 1^{4+}. TTF (3, Fig. 2) is one [71] of the strongest organic electron donors (E_{ox} = + 0.34 V) and exhibits [66] strong complexation with 1^{4+} (K_a = 6900 M^{-1}). Benzidine 5 (E_{ox} = + 0.59 V) [72], 1,5-dimethoxynaphthalene 6 (E_{ox} = + 1.11 V) [73], and 1,4-dimethoxybenzene 2 (E_{ox} = 1.31 V) [74] are all much weaker donors. Accordingly, a qualitative decrease in the K_a for the complex between 1^{4+} and benzidine 5 (K_a = 1044 M^{-1}) [75], 1,5-dihydroxynaphthalene 16 (Fig. 5) (K_a = 440 M^{-1}) [66], and 1,4-dimethoxybenzene 2 (K_a = 18 M^{-1}) has been observed [58].

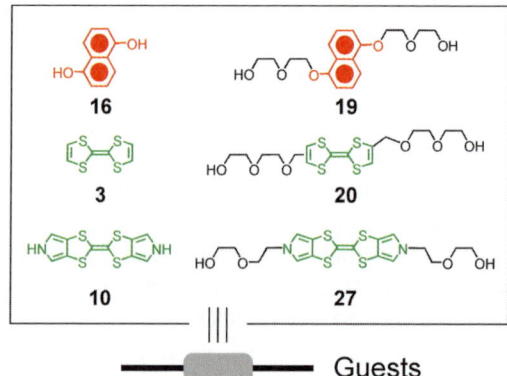

Fig. 5 Model π-electron-donating recognition units for the complexation with 1^{4+}

2.2
[C–H···π] and [C–H···O] Interactions

The host-guest complex [58] can be stabilized by [C–H···π] interactions [41–44] between the hydrogen atoms of the aromatic guest and the p-xylyl spacers of the 1^{4+} cyclophane host (see Fig. 6). Hydroquinone and DNP units both show close [C–H···π] contacts, with distances of ∼ 2.8 Å in addition to the π–π stacking (∼ 3.8 Å). Another important interaction arises from the [C–H···O] hydrogen bonding [46–54] between the α-protons of the bipyridinium rings in the cyclophane and the oxygen atoms of DEG chains that can be introduced on either side of the guest (∼ 3.2–3.3 Å, Fig. 6),

Towards a Rational Design of Molecular Switches and Sensors

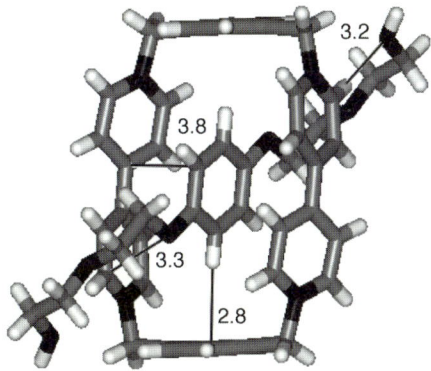

Fig. 6 X-ray crystal structure of the inclusion complex of a diethylene glycol substituted hydroquinone with 1^{4+}, showing $[\pi \cdots \pi]$, $[C-H \cdots \pi]$, and $[C-H \cdots O]$ interactions

thereby creating a rod-like guest. A study by Kaifer and co-workers [58] demonstrates that the addition of ethylene glycol arms onto hydroquinone results in upto a 200-fold increase in the binding constant (Table 1).

Table 1 Association constants of substituted hydroquinones measured by UV/Vis titration in MeCN[58]

R in R–⬡–R	K_a (M^{-1})
⟩O–Me	18
⟩O∼O–Me	290
⟩O∼O∼O–Me	3200
⟩O∼OH	340
⟩O∼O∼OH	3400

2.3
π-Surfaces

A study [71] on the complexation of different TTF derivatives with 1^{4+} has revealed that, in addition to the electron-donating strength, the size of the donor's π surface plays an important role. With a larger π-surface, the π–π stacking interactions [45] between the π-electron deficient bipyridinium units of the host and a π-electron rich guest are more efficient and lead to a more stable complex. Although the first oxidation potentials ($E_{ox} = +0.34$ V and $+0.38$, respectively) of TTF (**3**) and bis(pyrrolo)TTF (BPTTF, **10**) are approximately the same [71], **10** has a more extended π-surface. As a result, the association constant of **10** is [76] almost a factor of five higher (K_a = 12 000 M^{-1} compared to K_a = 2600 M^{-1} for **3**, MeCN). Similarly, a study [77] comparing the K_a (MeCN) of indole **7** and catechol **8** (Fig. 2) shows a higher K_a for **7**, which has a larger π-surface, but a lower first oxidation potential.

2.4
Solvent Effects

For the complexes between 1^{4+} and electron donors, higher binding affinities are observed [77] by going to more polar solvents. Complexation of the electron donors with the cyclophane is associated with a large decrease in the area of the aromatic surfaces of the donor exposed to solvent molecules. Polar solvents show, in general, low molecular polarizabilities and large cohesive forces [78]. Complex formation will therefore be favored in more polar solvents. A linear correlation between $-\Delta G_a$ of complexation and the solvent's Z-values [79] or E_T (30) [80] values has been observed [77].

3
Shuttling in Degenerate [2]Rotaxanes—Molecular Shuttles

A degenerate [2]rotaxane (molecular shuttle) is a system with two *identical* recognition sites encircled by one cyclophane. The potential energy surface is symmetric ($\Delta G = 0$). The more interesting parameter for molecular shuttles is therefore $\Delta G^{\ddagger}$, which determines the rate of shuttling between the two recognition sites.

3.1
Influence of K_a on Shuttling Rate

Figure 7 shows the PES of two degenerate systems with different recognition sites, but identical spacers [81]. Using these identical spacers the height of the transition states should be similar and $\Delta G^{\ddagger}$ is mainly determined by the location of the bottom of the wells, a quantity that is related (Fig. 3) to ΔG_a. A higher association constant K_a for the complexation between the model guests and the cyclophane will result in a lowering of the bottom of the ground state well (more negative ΔG_a) and therefore a larger $\Delta G^{\ddagger}$. Consequently, one expects that larger barriers (Fig. 7) such as $\Delta G^{\ddagger,2}$ will be obtained from molecular shuttles constructed with stronger recognition units (green).

This principle can be illustrated (Fig. 8) with the molecular shuttles 17^{4+} and 18^{4+} [81]. The [2]rotaxanes with the DNP units show a shuttling barrier of $\sim$ 15.1–15.5 kcal mol^{-1}, whereas those with the stronger π-electron donating TTF units have a $\Delta G^{\ddagger}$ value of $\sim$ 17–18 kcal mol^{-1}. The major portion of

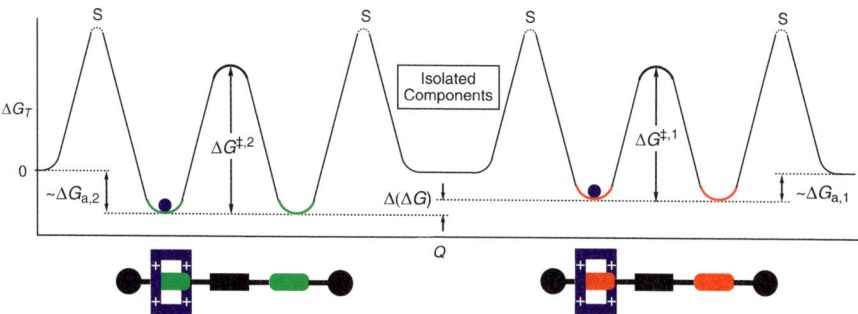

Fig. 7 Potential energy surface diagram of degenerate [2]rotaxanes. The energy barriers are defined with reference to the energy wells, the difference between each well is defined and the bottom of each well is referenced to the isolated components

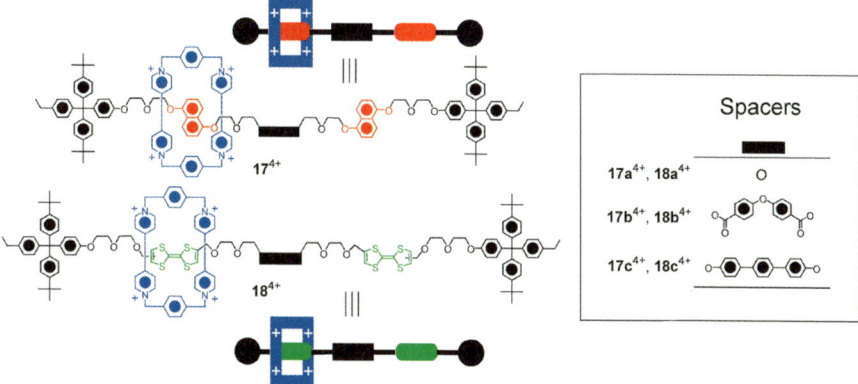

Fig. 8 Degenerate [2]rotaxanes 17^{4+} and 18^{4+} with different spacers

this difference in $\Delta G^{\ddagger}$ can be attributed to the differences in binding energies of a DNP model guest (**19**, $\Delta G_a = -6.3$ kcal mol^{-1}, MeCN, $T = 298$ K) and TTF model guest (**20** (*vide infra*), $\Delta G_a = -7.7$ kcal mol^{-1}, MeCN, $T = 298$ K) [66], each carrying two DEG chains (Fig. 5), with the cyclophane 1^{4+}.

3.2
Influence of the Spacers

The free energy barriers $\Delta G^{\ddagger}$ for the molecular shuttles $17a^{4+}$–$17c^{4+}$ are remarkably insensitive to the nature of the spacer. The differences between the free energy barriers $\Delta(\Delta G^{\ddagger})$ are smaller than 0.5 kcal mol^{-1}. However, a closer inspection (Table 2) of the thermodynamic parameters $\Delta H^{\ddagger}$ and $\Delta S^{\ddagger}$ reveals [81] the differences between the spacers.

The DEG unit (in $17a^{4+}$) which is the most flexible spacer, results in a larger loss in conformational flexibility in the transition state compared

Table 2 Activation parameters for the shuttling in the [2]rotaxanes **17a**$^{4+}$–**17c**$^{4+}$ in CD$_3$COCD$_3$

	$\Delta H^\ddagger$ (kcal mol^{-1})	$\Delta S^\ddagger$ (cal mol^{-1} K^{-1})	$\Delta G^\ddagger_{298}$ (kcal mol^{-1})^a
17a^{4+}	12.7	–9.3	15.5
17b^{4+}	14.5	–3.2	15.5
17c^{4+}	13.6	–4.8	15.1

^{a} ±0.1 kcal mol^{-1}

to the more rigid spacers of **17b**$^{4+}$ and **17c**$^{4+}$. This difference is reflected in the more negative $\Delta S^\ddagger$ value for **17a**$^{4+}$. The enthalpic contribution of the polyethylene glycol spacer in **17a**$^{4+}$ is smaller than for the larger and more rigid spacers in **17b**$^{4+}$ and **17c**$^{4+}$. This difference is believed to be caused by the more favorable electrostatic interactions between the electron-rich oxygen atoms and the electron-poor cyclophane as well as the larger flexibility. Additionally, the bent nature of the spacer in **17b**$^{4+}$ presumably causes some steric hindrance to the passage of the cyclophane, resulting in a larger enthalpic penalty.

An understanding of how to tune these thermodynamic parameters plays an important role in the design of molecular machines. For some materials applications, it is desirable that the molecular machine shows the same physical properties over a larger range of temperatures. For this purpose, a spacer with a small enthalpic contribution should be used in order to minimize the temperature dependent performance. Faster shuttling, for instance, can be achieved by choosing a spacer with a less negative entropic and a smaller enthalpic contribution.

4
Bistable [2]Rotaxanes—Molecular Switches

Although a large range of bistable rotaxanes have been synthesized [18, 57, 82, 83], only those composed of the weaker π-donor DNP unit with one of the stronger donor units TTF [66, 84–88], monopyrrolo TTF (MPTTF, **9**) [86–92], or BPTFF [66, 93] have been studied in detail. Therefore, we will limit our discussion to the [2]rotaxanes **21**$^{4+}$–**26**$^{4+}$ displayed in Fig. 9. For these bistable rotaxanes, the GSCC is defined as the translational isomer where the cyclophane resides on the TTF-based unit. In the MSCC, the cyclophane is located on the DNP unit.

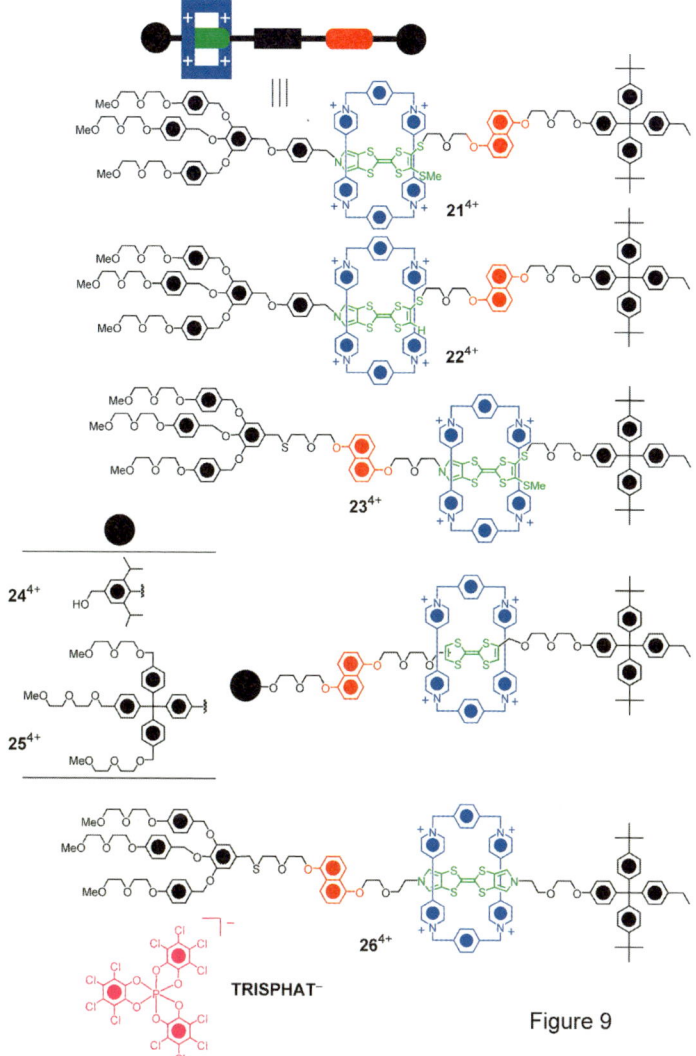

Fig. 9 Bistable [2]rotaxanes 21^{4+}–26^{4+} and structure of the TRISPHAT$^-$ counterion

4.1
Equilibrium Population Ratios

In the ideal case, a bistable [2]rotaxane will have a kind of PES shown in Fig. 4, with a relatively large ΔG, a situation which will result in a large population ratio ($N_{\text{GSCC}}/N_{\text{MSCC}}$) between the two translational isomers (ideally > 99 : 1). This ratio is important for the use of the bistable rotaxanes in device applications.

4.1.1
Solvent Effects

The bistable [2]rotaxanes 21^{4+}–24^{4+}, and 26^{4+} (Fig. 9) show very similar solvent effects (Table 3). In Me_2CO, the cyclophane has a high affinity for the DNP recognition site such that half of the population is in the MSCC. In more polar solvents, such as MeCN, the affinity shifts toward the TTF-based units. In such solvents, the electrostatic forces and [C–H···O] hydrogen bonds between the oxygen atoms of the side chains and the α C–H of the bipyridinium units in the cyclophane become less dominant because of competition with the solvent molecules. Comparing the K_a values between TTF (3) and the TTF guest 20 with two DEG chains (Fig. 5, Table 4) shows an increase in the binding affinity of 55-fold. Addition of two DEG chains to a DNP ring system (16 and 19, respectively) gives a more than 80-fold increase in K_a, suggesting that the influence of the DEG chains on the binding affinity is larger for DNP than it is for TTF. Therefore, decreasing the influence of [C–H···O] interactions in more polar solvents will favor the TTF-based guests [94].

Furthermore, in less polar solvents the tetracationic cyclophane will experience unfavorable interactions with the solvent. These interactions can be minimized by folding of the dumbbell [94], thereby placing the DNP inside the cavity of the cyclophane in order to utilize the larger π–π interactions between the TTF unit and the outer π-surface of the tetracation. In more polar solvents, the solvent-cyclophane interactions become less important and the ratio of the translational isomers should reflect [96] the relative π-donating capacities of the recognition units.

From the data in Table 3, it can be concluded that the ability to become involved in [C–H···O] interactions is an important factor in these bistable rotaxanes. The importance of this [C–H···O] interaction can be identified by the influence of the hydrophilic stopper in 21^{4+} and 23^{4+}. In 23^{4+}, the DNP unit is located next to the hydrophilic stopper and it is observed that the cyclophane favors the DNP unit over the MPTTF by a factor of 3 : 1 in Me_2CO. In 21^{4+}, the MPTTF is placed next to this hydrophilic stopper, resulting in an evening out of the ratio to 1 : 1 in Me_2CO. This balancing out is promoted by a folding over of the polyethylene glycol units on the hydrophilic stopper onto the outer surface of the tetracationic cyclophane.

Another indication for the importance of the [C–H···O] interactions between the cyclophane and the hydrophilic stopper is that for 21^{4+} and 22^{4+} similar population ratios are found in Me_2CO and Me_2SO. Although Me_2SO is a more polar solvent than MeCN, a smaller population ratio is observed. Going from MeCN to Me_2CO to Me_2SO, a decrease in the intensity of the absorption bands in the UV/Vis spectra was observed, suggesting a smaller N_{GSCC}/N_{MSCC} ratio in the presence of a weaker CT interaction between the π-electron donor recognition site and the π-electron-poor cyclophane aris-

Table 3 Population ratios (N_{GSCC}/N_{MSCC}) of bistable rotaxanes determined by NMR in different solvents at 298 K, unless otherwise stated

	Me$_2$CO	MeCN	Me$_2$SO
21^{4+}	1	3	1
$22a^{4+\,a}$	1.34	4	–[b]
$23a^{4+}$	0.33	1	2
$24^{4+\,c}$		~9	
$26^{4+\,d}$	1.2	4	

[a] Calculated by extrapolation of the plot of N_{GSCC}/N_{MSCC} to T [92]
[b] Not determined by NMR; UV/Vis spectroscopy suggests a similar value as for Me$_2$CO [92]
[c] Temperature 278 K [66]
[d] Temperature 295 K [93]

Table 4 Thermodynamic data at 298 K in MeCN for the binding of **3, 10, 16, 19, 20,** and **27**[a] with the tetracationic cyclophane and for the translational isomerism in the [2]rotaxanes 22^{4+}-24^{4+}, and 26^{4+} [b]

	ΔH (kcal mol^{-1})	ΔS (cal mol^{-1} K^{-1})	ΔG (kcal mol^{-1})	K_a (10^3 M^{-1})
16	– 16.04 ± 8.11	– 41.7	– 3.63 ± 0.36	0.44 ± 0.13
19	– 15.41 ± 0.02	– 30.8	– 6.26 ± 0.04	36.4 ± 0.25
3	– 10.64 ± 0.12	– 18.1	– 5.27 ± 0.03	6.9 ± 0.18
20	– 14.21 ± 0.06	– 22.1	– 7.66 ± 0.07	380.0 ± 22.0
10	– 9.00 ± 0.02	– 7.9	– 6.66 ± 0.03	70.8 ± 0.98
27	– 8.20 ± 1.70	– 3.6	– 7.17 ± 0.12	168.0 ± 17.0
22^{4+} (Me$_2$CO)	3.72 ± 0.34	14.2 ± 1.5	– 0.49	
22^{4+} (MeCN)	4.35 ± 0.03	18.5 ± 0.1	– 1.15	
23^{4+}	6.33 ± 0.18	21.0 ± 0.8	0.05	
$24^{4+\,c}$	2.82 ± 1.89	14.7 ± 6.8	– 1.56 ± 0.24	
26^{4+}	7.33 ± 0.36	28.6 ± 1.3	– 1.20 ± 0.03	

[a] Determined by ITC, ΔH corresponds to ΔH_a, ΔS to ΔS_a, ΔG to ΔG_a
[b] Determined by VT NMR for 22^{4+} and 23^{4+} and by VTT CV for 24^{4+} and 26^{4+} [66]
[c] The linear fit of $\Delta G/T$ vs. T^{-1} produces a low R value of 0.6, because ΔG for 24^{4+} was reasonably insensitive to temperature changes and therefore the data obtained reflects the standard error of the CV measurements

ing from an increase in the hydrogen-bonding between the solvent and the cyclophane. The decrease in intensity was larger for the absorption band assigned to the charge-transfer between the MPTTF unit next to the hydrophilic stopper and the cyclophane in 21^{4+} and 22^{4+}. Accordingly, in Me$_2$SO, a lower population ratio (N_{GSCC}/N_{MSCC}) was observed than in MeCN. In 23^{4+}, however, the DNP unit is located next to the hydrophilic stopper, thereby favoring the GSCC in Me$_2$SO.

4.1.2
Counterion Effects

The influence of the counterion has been studied by exchanging the commonly used PF_6^- counterion in the bistable rotaxane 26^{4+} for the more bulky TRISPHAT$^-$ counterion (Fig. 9). In Me$_2$CO, 45% of the population of the cyclophane resides on the DNP unit for $26 \cdot 4PF_6$, but for $26 \cdot 4$TRISPHAT less than 5% of the MSCC was found [93]. In the more polar solvent MeCN, the effect was less pronounced with a smaller change (25% to 15%) in the MSCC population upon replacement of PF_6^- or TRISPHAT$^-$. This observation can be explained by the fact that, in less polar solvents, the counterions are in closer proximity with the cyclophane and will have a profound influence on the noncovalent bonding interactions. In more polar solvents, the ions are better solvated, resulting in a smaller influence of the counterions.

The fact that the GSCC is more favored with the TRISPHAT$^-$ counterion than with PF_6^- can presumably be ascribed to steric and electronic effects. The more bulky TRISPHAT$^-$ anion requires a larger average distance between the positive charges of the cyclophane and the counterion. Thus, the tetracationic cyclophane becomes a better electron acceptor and favors the binding to the stronger π-electron-donating BPTFF unit. The bulkiness of the TRISPHAT$^-$ anion also causes a shielding of the cyclophane, thereby decreasing the [C – H $\cdots$ O] interactions between the DEG chains attached to the recognition unit with the cyclophane. Since these interactions are more important for the binding affinity of the DNP unit to the cyclophane than for the BPTFF unit (*vide infra*), the GSCC will be favored.

4.1.3
Thermodynamic Parameters

The thermodynamic parameters for the complexation of **3, 10, 16, 19, 20**, and **27** (Fig. 5) with the tetracationic cyclophane and bistable [2]rotaxanes 22^{4+} –24^{4+}, and 26^{4+} (Fig. 9) are collected in Table 4. The thermodynamic parameters for the π-electron-donor/1^{4+} complexes were determined by isothermal titration calorimetry (ITC). The data for the bistable [2]rotaxanes were obtained by variable temperature (VT) NMR or variable time- and temperature-cyclic voltammetry (VTT CV). The thermodynamic data provides explanations for the observed population ratios and the temperature sensitivity of these ratios.

Comparison of the data for the different model compounds shows that the DEG chains have a large influence on the magnitude of the association constants. For DNP and TTF units, containing DEG side arms (**19** and **20**, respectively), the K_a values are more than 50- and 80-fold higher than those for the model guests lacking these substituents, namely, **16** and **3**, respectively. The small differences in ΔH and ΔS between **27** and **10** suggest that the DEG chains have much less influence on the BPTTF unit: the association constant merely doubles upon introduction of the DEG substituents.

The larger increase in K_a at r.t. on the introduction of the DEG chains for the TTF model guest **20** compared to the BPTTF model guest **27** results in better population ratios for TTF- than for BPTTF-based [2]rotaxanes (24^{4+} vs. 26^{4+}, Table 3). An MPTTF unit can be regarded as a combination of a TTF and a BPTTF unit. The MPTTF-based [2]rotaxane 23^{4+} reveals thermodynamic parameters which are somewhere in between the values for the TTF-based rotaxane 24^{4+} and BPTTF-based rotaxane 26^{4+}.

Interestingly, although the addition of DEG chains enhances the binding affinity of all guests by the tetracationic cyclophane, they impact ΔH and ΔS for every unit differently. For TTF, the DEG chains cause a better enthalpy (favoring binding) but a worse entropy. For DNP and BPTTF the opposite effect is observed with a slightly worse enthalpy and a better entropy.

The van 't Hoff plot (Fig. 10) of $R \ln (N_{GSCC}/N_{MSCC})$ against $1000\,T^{-1}$ summarizes the properties of the different rotaxanes. The slope of the plot corresponds to $-\Delta H$ and the intercept with the y-axis gives the value for ΔS (Eq. 3). A larger slope, and therefore a larger ΔH, correlates to a more temperature-sensitive population ratio.

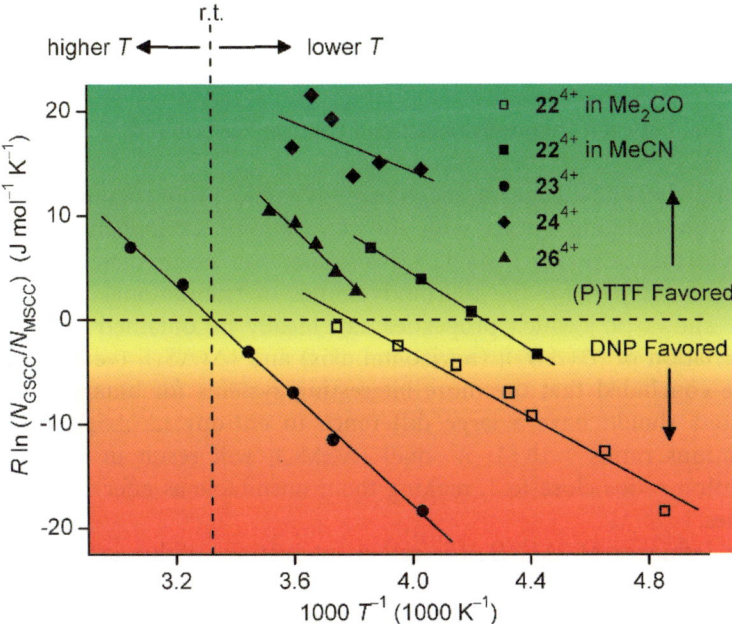

Fig. 10 Van 't Hoff plot of bistable [2]rotaxanes 22^{4+}–24^{4+} and 26^{4+} in MeCN. The background color corresponds to the population ratio—pure green is > 10 : 1 GSCC:MSCC and pure red is < 1 : 10

For all (P)TTF/DNP-based [2]rotaxanes, the MSCC is favored at low temperatures because of the positive ΔH value. The positive ΔS favors the GSCC at higher temperatures, since the ΔS value will become more dominant than the $\Delta H/T$ factor, which becomes smaller at higher temperatures. Figure 10 also shows that with a larger ΔS value, better population ratios in favor of the GSCC are obtained.

Since ΔH and ΔS represent the enthalpy and entropy changes between the two recognition units in the [2]rotaxanes, these values can be directly related to the difference in enthalpy and entropy changes, $\Delta(\Delta H_a)$ and $\Delta(\Delta S_a)$ for the single components (Table 4). The enthalpy difference $\Delta(\Delta H_{a,(P)TTF} - \Delta H_{a,DNP})$ and the entropy difference $\Delta(\Delta S_{a,(P)TTF} - \Delta S_{a,DNP})$ give values of $\Delta(\Delta H_a) = 1.2$ kcal mol^{-1} and $\Delta(\Delta S_a) = 8.7$ cal mol^{-1} K^{-1} for the guests **20** and **19**. By contrast, for **27** and **19**, the differences are higher: $\Delta(\Delta H_a) = 7.2$ kcal mol^{-1} and $\Delta(\Delta S_a) = 27.2$ cal mol^{-1} K^{-1}. These $\Delta(\Delta H_a)$ and $\Delta(\Delta S_a)$ values correspond well with the trends observed for ΔH and ΔS for the TTF- and BPTTF-based [2]rotaxanes **24**$^{4+}$ and **26**$^{4+}$. As a consequence, the bistable [2]rotaxane **26**$^{4+}$ shows a larger temperature sensitivity than **24**$^{4+}$, an outcome which is caused by the larger enthalpy difference ΔH. Comparing between the plots (Fig. 10) of **22**$^{4+}$ in Me$_2$CO and Me$_2$CN, suggests that the solvent's effect on the population ratio has both an enthalpic and entropic contribution. The MPTTF-based [2]rotaxanes **21**$^{4+}$–**23**$^{4+}$ show ΔH and ΔS values that lie between those of the TTF- and BPTTF-based [2]rotaxanes **24**$^{4+}$ and **26**$^{4+}$.

4.1.4
Correlation between Thermodynamics and Physical Performance

Applying the knowledge of the thermodynamic properties of the model π-electron-donating guests with the tetracationic cyclophane, the physical performance of the bistable [2]rotaxanes can be predicted and tuned. The quadrant plot for the thermodynamics of translational isomerism in Fig. 11 shows [97] which properties can ideally be obtained by a systematic variation of $|\Delta(\Delta H_a)|$ (horizontal axis) and $|\Delta(\Delta S_a)|$ (vertical axis). It can be concluded that the more interesting systems for binary device applications should have a large difference in entropy $|\Delta(\Delta S_a)|$ leading to populations ratios $>$ 10:1. A small $|\Delta(\Delta S_a)|$ will result in systems with population ratios close to 1, making them unsuitable as effective molecular switches.

The influence of $|\Delta(\Delta H_a)|$ is rather interesting. With a small $|\Delta(\Delta H_a)|$ and large $|\Delta(\Delta S_a)|$ (top left hand quadrant), the bistable [2]rotaxanes will show very stable and very good population ratios over a large temperature range, leading to reproducible behavior for materials applications. However, with a large $|\Delta(\Delta H_a)|$ and large $|\Delta(\Delta S_a)|$ (top right hand quadrant), the bistable [2]rotaxane will show a large temperature sensitivity and a com-

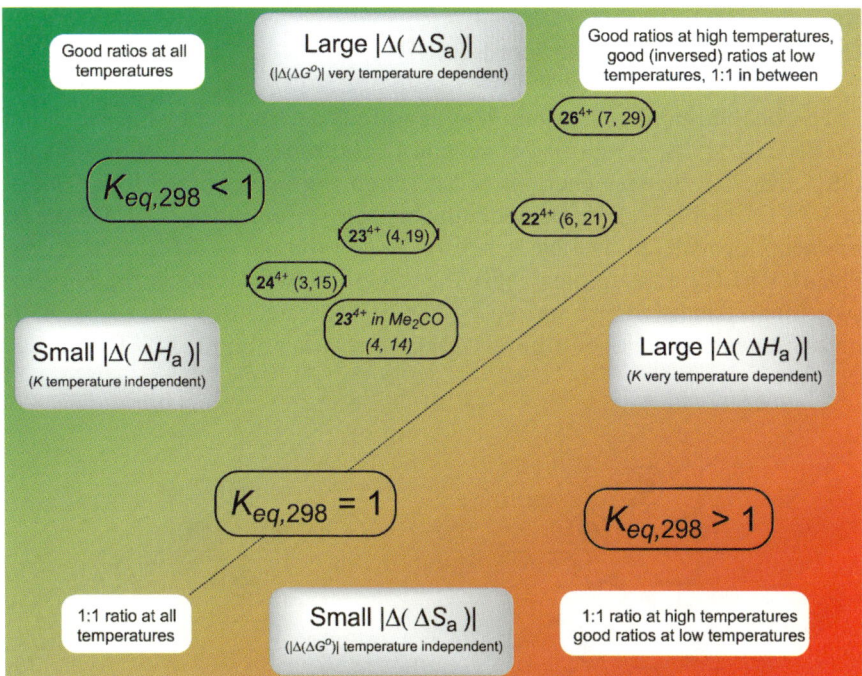

Fig. 11 Generic quadrant plot showing the influence of $|\Delta(\Delta H_a)|$ and $|\Delta(\Delta S_a)|$ of model recognition units on the physical performance of [2]rotaxanes containing these units. Overlaid are specific values for bistable [2]rotaxanes 22^{4+}–24^{4+}, and 26^{4+} in MeCN (ΔH in kcal mol^{-1} K^{-1}) and the color scheme corresponds to the population ratio distributions (K_{eq}) at 298 K. At higher temperatures the slope of the line bisecting the two colored sections becomes shallower

plete inversion of a population ratio can be obtained within a relatively small temperature range. These bistable [2]rotaxanes should be good temperature sensors.

The available bistable [2]rotaxanes 22^{4+}–24^{4+}, and 26^{4+} are located rather high in the quadrant plot. The TTF/DNP-based [2]rotaxane 24^{4+} shows good population ratios (9 : 1) and is rather temperature insensitive. By a further increase of $|\Delta(\Delta S_a)|$ and decrease of $|\Delta(\Delta H_a)|$, even better switches could be obtained. The BPTTF/DNP-based [2]rotaxane 26^{4+} is located in the upper right corner. A major challenge stemming from these observations is to find new and better combinations of recognition units, which can be incorporated into bistable [2]rotaxanes, so that their switching properties can be even further improved. Given the correlation between the population ratios and the properties of the model host-guest complexes, comprehensive studies on new model guests should provide a rational way to design bistable rotaxanes from their component building blocks.

4.2
Chemical Tuning of Switching Speeds ($\Delta G^{\ddagger}$)

The switching speed at a certain temperature is determined by the activation barrier $\Delta G^{\ddagger}$ (Eqs. 4 and 5). For several [2]rotaxanes and a pseudorotaxane (28^{4+}, Fig. 12), the free energies of activation were determined by VT NMR or VTT CV (Table 5). For fast switching bistable [2]rotaxanes, VT NMR measurements provide an average value for $\Delta G^{\ddagger}$, (average of $\Delta G^{\ddagger,1}$ and $\Delta G^{\ddagger,2}$, Fig. 4). VTT CV measurements give the values for $\Delta G^{\ddagger,1}$ (*vide infra*). For slow switching systems, $\Delta G^{\ddagger,1}$ and $\Delta G^{\ddagger,2}$ can be separately determined by NMR measurements, by following the relaxation from either the pure MSCC or GSCC.

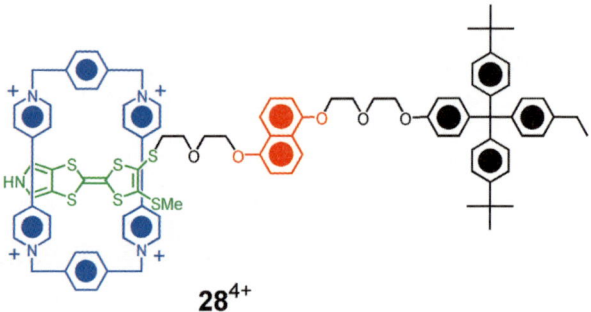

Fig. 12 Bistable [2]pseudorotaxane 28^{4+}

Table 5 Free energies of activation, enthalpy, and entropy for the switching process in [2]rotaxanes and pseudorotaxanes at 298 K, determined by NMR in Me$_2$CO, unless otherwise stated. $\Delta G^{\ddagger}$ is $\Delta G^{\ddagger,1}$, unless otherwise stated

	$\Delta H^{\ddagger}$ (kcal mol^{-1})	$\Delta S^{\ddagger}$ (cal mol^{-1} K^{-1})	$\Delta G^{\ddagger}$ (kcal mol^{-1})
21^{4+}	17	−24	24 [a][90,92]
21^{4+} (Me$_2$SO)			22 [a,b][90]
22^{4+}			14.8 [c][92]
22^{4+} (MeCN)			14.7 [c][92]
28^{4+} [d]	17.0	−23	23.8 [91]
28^{4+} [d]	16.4	−24	23.6 [b][91]
28^{4+} [d] (MeCN)	19.8	−12	23.8 [91]
24^{4+}	12.2	−13.6	16.2 ± 0.3 [e][66]
26^{4+}	6.4 ± 0.5	−37.8 ± 1.8	17.66 ± 0.05 [e][66]

[a] ±0.1 kcal mol^{-1}
[b] $\Delta G^{\ddagger,2}$
[c] Temperature 293 K, $\Delta G^{\ddagger,\text{average}}$
[d] Estimated errors are ± 5% for $\Delta G^{\ddagger}$, ± 5% for $\Delta H^{\ddagger}$, and ± 25% for $\Delta S^{\ddagger}$
[e] Determined by VTT CV

In Me$_2$CO and Me$_2$SO, $\Delta G^{\ddag,1}$ should be equal to $\Delta G^{\ddag,2}$ for 21^{4+}, since the ratio of the co-conformations is 1 : 1 in these solvents, resulting in $\Delta G = 0$ at 298 K. The model complex for 21^{4+}, pseudorotaxane 28^{4+}, shows indeed $\Delta G^{\ddag,1} = \Delta G^{\ddag,2}$ in Me$_2$CO. Unfortunately, only the value for $\Delta G^{\ddag,1}$ has been reported for 21^{4+} in MeCN. It would be interesting to find out if the value for $\Delta G^{\ddag,2}$ would be larger than for $\Delta G^{\ddag,1}$, as would be expected, since $N_{GSCC}/N_{MSCC} = 3$.

Pseudorotaxane 28^{4+} shows a more negative entropy of activation $\Delta S^{\ddag}$ in Me$_2$CO than in MeCN. This fact can possibly be explained by the higher solubility of 28^{4+} in Me$_2$CO and by the lower polarity of Me$_2$CO. Because of the lower polarity of Me$_2$CO, the solvent molecules are more ordered around the aromatic surfaces in the transition state than in the ground state. Thus, a more negative activation entropy in Me$_2$CO than in MeCN could be obtained.

The free energies of activation $\Delta G^{\ddag}$ for rotaxanes 22^{4+} and 24^{4+} lie within the same range and are relatively close to the $\Delta G^{\ddag}$ value (15.5 kcal mol^{-1}) found for shuttling in the degenerate DNP-based [2]rotaxane $17a^{4+}$ (Fig. 8). The slightly higher value for $\Delta G^{\ddag}$ (17.7 kcal mol^{-1}) for 26^{4+} possibly arises from the lowering of the bottom of the ground-state well for the DNP unit, caused by the stabilizing interaction between the hydrophilic stopper and the cyclophane encircled DNP unit, a phenomenon that results in an increase in $\Delta G^{\ddag}$.

A large increase in the free energy barrier of $\sim$ 9 kcal mol^{-1} was obtained by the replacement of the hydrogen atom at the TTF side of the MPTTF unit in 21^{4+} by a methylthio group to give 22^{4+}. Accordingly, 21^{4+} shows [92] a much slower switching than 22^{4+}. The switching process in 21^{4+} is slowed down so much that the GSCC and MSCC could be separated chromatographically and isolated. In Me$_2$SO, the activation barrier is 2 kcal mol^{-1} lower than in Me$_2$CO ($\Delta G^{\ddag}$ = 22 kcal mol^{-1} vs. 24 kcal mol^{-1}), which is probably a direct consequence of the overall decrease in noncovalent bonding interactions, which raise the bottom of the ground-state wells, thereby lowering the activation barrier $\Delta G^{\ddag}$.

A similar change in kinetics, from fast to slow exchange, has been observed [71] by the introduction of SMe substituents on TTF (**3**) or on the pyrrole ring of BPTFF (**10**). The [2]rotaxane 23^{4+}, in which the SMe group of the MPTTF unit is oriented toward the stopper and not toward the spacer, exhibits fast switching (equilibrium is reached within 5 min. at r.t.). These observations constitute additional proof that the SMe group causes the large activation barrier $\Delta G^{\ddag}$ and therefore slow switching in 21^{4+}.

Moving from a methylthio substituent to the even bulkier ethylthio group on a MPTTF unit slows the exchange down further. For a MPTTF-SEt/1^{4+} complex, no dethreading/exchange can be observed [66, 90] at temperatures up to 425 K.

4.3
Switching in Bistable Rotaxanes

Besides the spontaneous movement of the cyclophane between two recognition units in a (P)TTF/DNP-based [2]rotaxane, a forced translation to the DNP unit can be accomplished by the oxidation of the TTF unit. The TTF$^{·+}$ radical cation, experiences a strong electrostatic repulsion from the tetracationic cyclophane, a situation which results in an increase in free energy (ΔG_T) of the bottom for the well of this recognition unit (Scheme 4). Accordingly, the cyclophane moves to the DNP unit. After reduction of the TTF$^{·+}$ radical cation the cyclophane moves back to the more favored TTF unit. However, this movement is thermally-activated and thus a metastable state is formed in which the tetracationic cyclophane resides for a detectable period of time on the DNP unit before moving back to the TTF unit. This slow relaxation to the ground state can be followed by VTT CV measurements. In

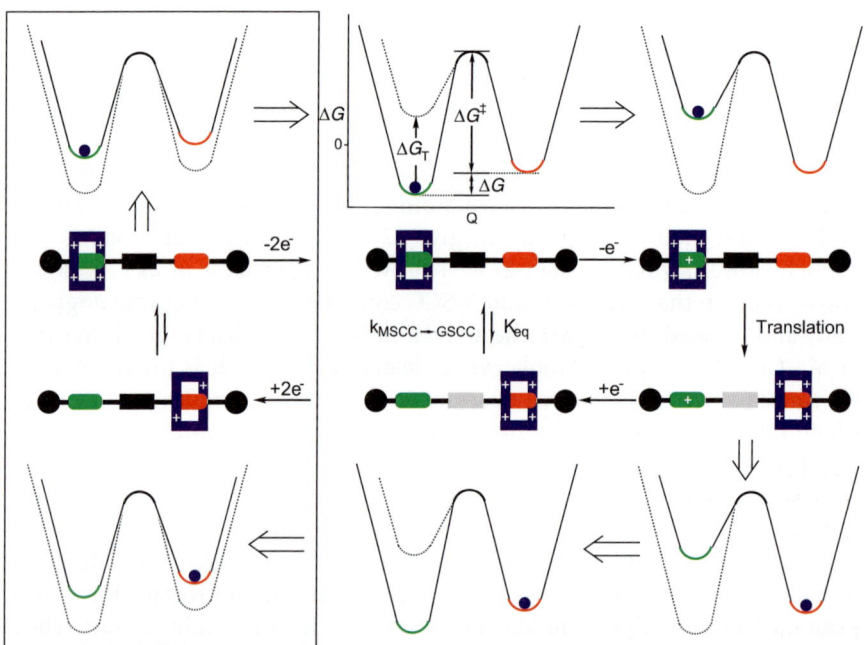

Scheme 4 Schematic representation and associated potential energy surface (PES) diagrams for the switching cycle of "forced" translation of the cyclophane (*blue*) from the TTF unit (*green*) to the DNP unit (*red*). Oxidation produces a different shape to the PES that favors energetically the DNP site. Inset: Proposed PES diagrams associated with the reduction showing how the barrier gets smaller while the TTF unit is still favored on account of its strong donor ability in the face of the now weaker diactionic cyclophane acceptor

addition to VT NMR, this method provides yet another tool for the determination of the energy barrier $\Delta G^{\ddagger,1}$ from the MSCC to the GSCC. In a process that is not fully understood, the slow relaxation can be sped up by the reduction of the cyclophane to its diradical form, $CBPQT^{2\bullet/2+}$. Presumably the barrier to movement is lowered, concomitant with a decrease in the noncovalent interactions with the DNP and neutral TTF units. Both effects cause the translation to the TTF unit and the subsequent re-oxidation of the cyclophane (Scheme 4) reinstating the GSCC.

5
Environmental Effects

The thermodynamic and kinetic behavior of the [2]rotaxanes is dependent on their physical environment. The recovery of the GSCC from the MSCC can be measured [16, 66] in solution (MeCN) [66], in a liquid polymer matrix [85], in a self-assembled monolayer (SAM) on gold [84], and in molecular-switch tunnel junctions (MSTJs). The ΔG and $\Delta G^{\ddagger}$ values for 24^{4+} and 26^{4+} in solution and in the polymer matrix were determined by VTT CV measurements [66]. For the MSTJ device, a temperature-dependent switching amplitude (I_{OPEN}/I_{CLOSED}) was detected for 25^{4+} and 26^{4+}, where 25^{4+} is the amphiphilic analog of 24^{4+}. On the basis of the proposed switching mechanism of bistable rotaxanes in MSTJ devices, I_{OPEN}/I_{CLOSED} represents the ratio N_{MSCC}/N_{TOTAL} of the bistable rotaxanes measured in solution and polymer phases at different stages throughout the switching cycle [66].

Figure 13 shows the temperature dependence of the GSCC population at equilibrium. In solution, polymer, and MSTJ, the percentage of GSCC for the TTF-based [2]rotaxanes 24^{4+} and 25^{4+} is stable remaining at around 85–90% over a large temperature range. The percentage of GSCC varies more heavily for 26^{4+} from $\sim 50\%$ to $\sim 80\%$ upon an increase in temperature. The fact that the trends in the temperature dependence of the GSCC population for these bistable [2]rotaxanes is maintained in all three environments, suggests that the GSCC population is dictated by the chemical structure of the [2]rotaxanes, rather than by the physical environment of the molecules.

Although the influence of the environment on the separate thermodynamic parameters (Tables 6 and 7) is not yet understood, it can be observed that, for both the TTF- and BPTTF-based [2]rotaxanes, the free energy difference between the GSCC and MSCC, ΔG_{298} decreases in magnitude upon moving from solution to polymer to MSTJ. This decrease in ΔG correlates with a preference of the MSCC over the GSCC, by going from solution to polymer to MSTJ (see Fig. 13). The effect is less pronounced for the TTF-based rotaxanes where it appears that the similarity observed for the enthalpy and entropy of the model guests, **19** and **20**, leads to the modest effect across the three environments. However, the large thermodynamic differences between

Table 6 Thermodynamic parameters of 24^{4+} and 25^{4+} in different environments [66]

Environment	ΔH (kcal mol^{-1})	ΔS (cal mol^{-1} K^{-1})	ΔG_{298} (kcal mol^{-1})
24^{4+} Solution	2.82 ± 1.89	14.7 ± 6.8	−1.56 ± 0.24
24^{4+} Polymer	2.35 ± 0.47	12.3 ± 1.6	−1.32
25^{4+} MSTJ	2.90 ± 0.98	13.2 ± 3.1	−1.03

Table 7 Thermodynamic parameters of 26^{4+} in different environments [66]

Environment	ΔH (kcal mol^{-1})	ΔS (cal mol^{-1} K^{-1})	ΔG_{298} (kcal mol^{-1})
Solution	7.33 ± 0.36	28.6 ± 1.3	−1.20 ± 0.03
Polymer	6.16 ± 0.23	21.4 ± 0.8	−0.22
MSTJ	4.69 ± 0.29	15.9 ± 1.0	−0.05

Table 8 Kinetics data for the relaxation of the MSCC to the GSCC for 26^{4+} and $\Delta G^{\ddagger}_{298}$ for 24^{4+} and 25^{4+} [66]

Environment	$\Delta H^{\ddagger}$ (kcal mol^{-1})	$\Delta S^{\ddagger}$ (cal mol^{-1} K^{-1})	$\Delta G^{\ddagger}_{298}$ (kcal mol^{-1})	$\Delta G^{\ddagger}_{298}$ (kcal mol^{-1}) (24^{4+})	$\Delta G^{\ddagger}_{298}$ (kcal mol^{-1}) (25^{4+})
Solution	6.4 ± 0.5	−37.8 ± 1.8	17.66 ± 0.05	16.2 ± 0.3	
Polymer	8.6 ± 1.1	−34.3 ± 3.7	18.85 ± 0.01	18.1 ± 0.2 [a]	
MSTJ	14.1 ± 1.4	−23.9 ± 4.6	21.26 ± 0.04		22.21 ± 0.04

[a] Value for relaxation free energy barrier 24^{4+} in SAM $\Delta G^{\ddagger}_{298} = 18.0c \pm 0.5$ kcal mol^{-1}

the model guests, **19** and **27**, for the bistable rotaxane 26^{4+} are amplified such that for the BPTTF-based rotaxane the impact of the environment is large enough that in MSTJs the switching ratio has almost evened out to be 1 : 1.

Figure 14 shows the relaxation kinetics from the MSCC to the GSCC of the TTF- and BPTTF-based [2]rotaxanes. The kinetic data in Table 8 is derived from this plot according to the Eyring equation (Eq. 5). Going from solution to MSTJ, the free energy of activation $\Delta G^{\ddagger}_{298}$ increases. For 26^{4+}, the free energy $\Delta G^{\ddagger}_{298}$ increases from 17.7 to 18.9 to 21.3 kcal mol^{-1} on going from MeCN solution to polymer gels to MSTJs. For 24^{4+} and 25^{4+} the situation is qualitatively similar. This increase in the activation barrier is directly detectable, since the relaxation time increases for 26^{2+} from ∼ 1.5 s in solution to ∼ 10 s in the polymer to ∼ 630 s in the MSTJ at r.t.

A polymer matrix can be considered as a solvent with a very high viscosity, which has been proven to slow down the rates of unimolecular reaction of small molecules [98–102] and macromolecules [103, 104]. In the MSTJs, each

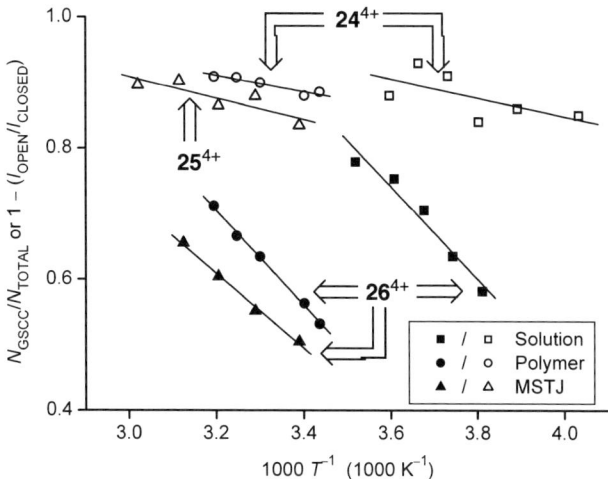

Fig. 13 Percentage GSCC at equilibrium for 24^{4+}–26^{4+} at different temperatures in solution, polymer matrix or MSTJ

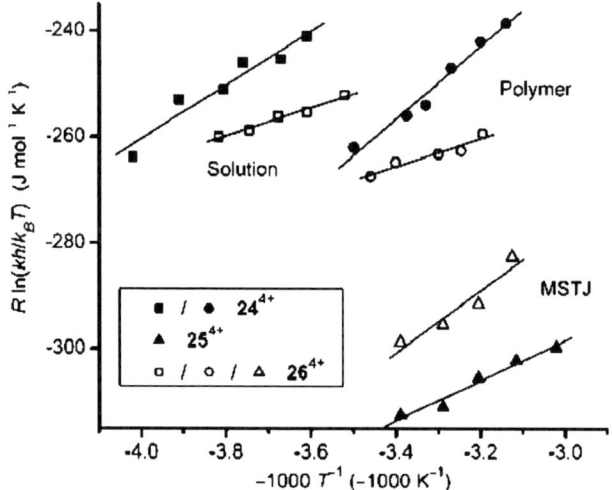

Fig. 14 Eyring plot for 24^{4+}–26^{4+} in different environments, solution, polymer matrix or MSTJ

bistable switch is closely-packed with the next, leading to a situation where nearest-neighbor intermolecular interactions influence the thermodynamics of activation.

Since both the thermodynamics and kinetics data show similar trends in all three environments, it can be concluded that the (forced) translation of the tetracationic cyclophane along the dumbbell in these bistable [2]rotaxanes can be assigned to a single mechanism for all environments. It is universal.

6
Conclusions

This Chapter demonstrates that the thermodynamics and kinetics—and therefore physical properties—of bistable [2]rotaxanes are influenced by a considerable collection of different factors, which include the nature of the recognition units, solvent, substituent effects, counterions, and physical environment. A basic understanding of how each factor influences the entropy and enthalpy of the bistable [2]rotaxanes will contribute to the design of improved molecular switches and machines.

We have shown that the equilibrium thermodynamical properties (ΔH and ΔS) of degenerate and bistable [2]rotaxanes are mainly determined by the difference in $\Delta(\Delta H_a)$ and $\Delta(\Delta S_a)$ of the corresponding model recognition units, studied as 1 : 1 complexes in [2]pseudorotaxanes with the tetracationic cyclophane as host. This feature provides us with the opportunity to design new [2]rotaxanes with improved physical performance, based on the knowledge of the thermodynamic properties of complexes of π-electron-donating guests with the tetracationic cyclophane. The more interesting bistable [2]rotaxanes for material applications should show a large difference in entropy ΔS. Dependent on the application either a very small ΔH (temperature stable population ratio) or very high ΔH (temperature sensitive ratio) is desirable.

The rate of shuttling between two stations can be influenced markedly by substituent effects. Placing a bulky substituent on a recognition unit can dramatically slow down and even stop the movement of the cyclophane along the dumbbell component of the [2]rotaxane. This movement is also slowed down by incorporating the bistable [2]rotaxane into a polymer matrix or into a MSTJ.

The investigations in different physical environments show that a single mechanism exists for the (forced) translation of the cyclophane along the dumbbell in bistable [2]rotaxanes, and it is independent of the environment.

The major challenge for the future is to specify which properties are desirable for a particular materials application and to use this recently accrued package of knowledge on thermodynamics and kinetics to design rationally the donor/acceptor bistable [2]rotaxane to match the application.

7
Epilogue

By using the knowledge provided in Fig. 11 and the thermodynamic data from Table 4, new and improved bistable [2]rotaxanes for materials applications can be designed. For example, one proposal (Fig. 15) for a thermally sensitive molecular switch is the bistable [2]rotaxane **29^{4+}**.

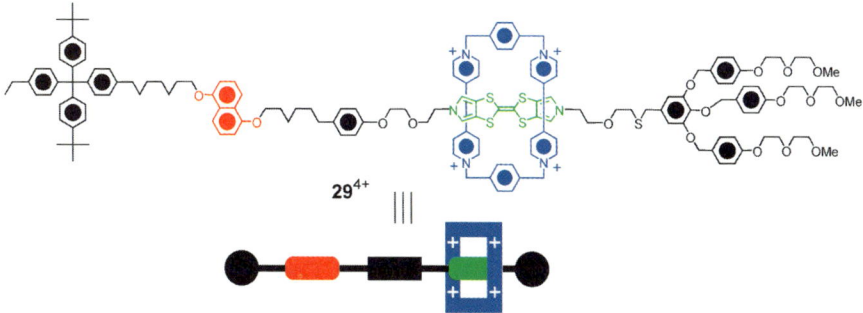

Fig. 15 Proposed bistable [2]rotaxane 29^{4+}

In order to make a switch that is located higher up in the right corner of the quadrant in Fig. 11 than **26**$^{4+}$, we propose the bistable [2]rotaxane **29**$^{4+}$. The model recognition units for **26**$^{4+}$—namely, guests **27** and **19**—show values of $\Delta(\Delta H_a) = 7.21$ kcal mol^{-1} and $\Delta(\Delta S_a) = 27.2$ cal mol^{-1} K^{-1}, respectively. The combination of **27** and **16** provides values of $\Delta(\Delta H_a) = 7.86$ kcal mol^{-1} and $\Delta(\Delta S_a) = 38.1$ cal mol^{-1} K^{-1}. Incorporating these two units into bistable [2]rotaxane **29**$^{4+}$ should result in a slightly more temperature sensitive switch, which should exhibit much better population ratios than **26**$^{4+}$. This bistable [2]rotaxane would be a good temperature sensor.

Acknowledgements This research was funded by the National Science Foundation (NSF), the Molectronics Program of the Defense Advanced Research Project Agency (DARPA), the Microelectronics Advance Research Corporation (MARCO) and its Focus Center on Functional Engineered NanoArchitectonics (FENA) and the Center for Nanoscale Innovation for Defense (CNID) in the US. NNPM acknowledges the Netherlands Organization for Scientific Research (NWO) for a TALENT fellowship.

References

1. Mendes PM, Flood AH, Stoddart JF (2005) Appl Phys A 80:1197
2. Huang TJ, Brough B, Ho CM, Liu Y, Flood AH, Bonvallet PA, Tseng HR, Stoddart JF, Baller M, Magonov S (2004) Appl Phys Lett 85:5391
3. Liu Y, Flood AH, Bonvallet PA, Vignon SA, Northrop BH, Tseng HR, Jeppesen JO, Huang TJ, Brough B, Baller M, Magonov S, Solares SD, Goddard WA, Ho CM, Stoddart JF (2005) J Am Chem Soc 127:9745
4. Nikitin K, Long B, Fitzmaurice D (2003) Chem Commun 282
5. Long B, Nikitin K, Fitzmaurice D (2003) J Am Chem Soc 125:5152
6. Nikitin K, Fitzmaurice D (2005) J Am Chem Soc 127:8067
7. Van Delden RA, Van Gelder MB, Huck NPM, Feringa BL (2003) Adv Funct Mater 13:319
8. Van Delden RA, Mecca T, Rosini C, Feringa BL (2004) Chem Eur J 10:61
9. Jiménez MC, Dietrich-Buchecker C, Sauvage J (2000) Angew Chem Int Ed 39:3284

10. Jiménez-Molero MC, Dietrich-Buchecker C, Sauvage JP (2002) Chem Eur J 8:1456
11. Jiménez-Molero MC, Dietrich-Buchecker C, Sauvage JP (2003) Chem Commun 1613
12. Sauvage JP (2005) Chem Commun 1507
13. Hernandez R, Tseng HR, Wong JW, Stoddart JF, Zink JI (2004) J Am Chem Soc 126:3370
14. Katz E, Lioubashevsky O, Willner I (2004) J Am Chem Soc 126:15520
15. Katz E, Sheeney-Haj-Ichia L, Willner I (2004) Angew Chem Int Ed 43:3292
16. Luo Y, Collier CP, Jeppesen JO, Nielsen KA, De Ionno E, Ho G, Perkins J, Tseng HR, Yamamoto T, Stoddart JF, Heath JR (2002) ChemPhysChem 3:519
17. Huang TJ, Tseng HR, Sha L, Lu W, Brough B, Flood AH, Yu BD, Celestre PC, Chang JP, Stoddart JF, Ho CM (2004) Nano Lett 4:2065
18. Flood AH, Liu Y, Stoddart JF (2004) From cyclophanes to molecular machines. In: Hopf H, Gleiter R (eds) Modern Cyclophane Chemistry. Wiley, Weinheim, p 485
19. Flood AH, Ramirez RJA, Deng WQ, Muller RP, Goddard III WA, Stoddart JF (2004) Aust J Chem 57:301
20. Stoddart JF, Tseng HR (2003) PNAS 99:4797
21. Doddi G, Ercolani G, Mencarelli P, Piermattei A (2005) J Org Chem 70:3761
22. Diederich F, Stang PJ (eds) (2000) Templated Organic Synthesis. Wiley, Weinheim
23. Aricó F, Badjic JD, Cantrill SJ, Flood AH, Leung KCF, Liu Y, Stoddart JF (2005) Top Curr Chem 249:203
24. Koumura N, Geertsema EM, Van Gelder MB, Meetsma A, Feringa BL (2002) J Am Chem Soc 124:5037
25. Bottari G, Dehez F, Leigh DA, Nash PJ, Pérez EM, Wong JKY, Zerbetto F (2003) Angew Chem Int Ed 42:5886
26. Leigh DA, Wong JKY, Dehez F, Zerbetto F (2003) Nature 424:174
27. Hernández JV, Kay ER, Leigh DA (2004) Science 306:1532
28. Albrecht-Gary AM, Dietrich-Buchecker C, Saad Z, Sauvage JP (1992) Chem Commun 280
29. Leigh DA, Troisi A, Zerbetto F (2000) Angew Chem Int Ed 39:350
30. Leigh DA, Troisi A, Zerbetto F (2001) Chem Eur J 7:1450
31. Odell B, Reddington MV, Slawin AMZ, Spencer N, Stoddart JF, Williams DJ (1988) Angew Chem Int Ed Engl 27:1547
32. Lu T, Zhang L, Gokel GW, Kaifer AE (1993) J Am Chem Soc 115:2542
33. Capobianchi S, Doddi G, Ercolani G, Keyes JW, Mencarelli P (1997) J Org Chem 62:7015
34. Kharitonov AB, Shipway AN, Katz E, Willner I (1999) Rev Anal Chem 18:255
35. Lahav M, Gabai R, Shipway AN, Willner I (1999) Chem Commun 1937
36. Lahav M, Shipway AN, Willner I (1999) J Chem Soc, Perkin Trans II 1925
37. Kaminski GA, Jorgensen WL (1999) J Chem Soc, Perkin Trans II 2365
38. Damgaard D, Nielsen MB, Lau J, Jensen KB, Zubarev R, Levillain E, Becher J (2000) J Mater Chem 10:2249
39. Zhang KC, Liu L, Mu TW, Guo QX (2001) Chem Phys Lett 333:195
40. Grubert L, Jacobi D, Buck K, Abraham W, Mügge C, Krause E (2001) Eur J Org Chem 20:3921
41. Hunter CA, Sanders JKM (1990) J Am Chem Soc 112:5525
42. Hunter CA (1993) Angew Chem Int Ed Engl 32:1584
43. Cozzi F, Siegel JS (1995) Pure Appl Chem 67:683
44. Shetty AS, Zhang J, Moore JS (1996) J Am Chem Soc 118:1019
45. Claessens CG, Stoddart JF (1997) J Phys Org Chem 10:254
46. Desiraju GR (1991) Acc Chem Res 24:290

47. Desiraju GR (1996) Acc Chem Res 29:441
48. Krishnamohan Sharma CV, Zaworotko MJ (1996) Chem Commun 2655
49. Steiner T (1997) Chem Commun 727
50. Berger I, Egli M (1997) Chem Eur J 3:1400
51. Bodige SG, Rogers RD, Blackstock SC (1997) Chem Commun 1669
52. Desiraju GR (1998) Top Curr Chem 198:57
53. Raymo FM, Bartberger MD, Houk KN, Stoddart JF (2001) 123:9264
54. Desiraju GR (2002) Acc Chem Res 35:565
55. Ashton PR, Odell B, Reddington MV, Slawin AMZ, Stoddart JF, Williams DJ (1988) Angew Chem Int Ed Engl 27:1550
56. Ashton PR, Goodnow TT, Kaifer AE, Reddington MW, Slawin AMZ, Spencer N, Stoddart JF, Vicent C, Williams DJ (1989) Angew Chem Int Ed Engl 28:1396
57. Raymo FM, Stoddart JF (2001) Switchable Catenanes, Molecular Shuttles. In: Feringa BL (ed) Molecular Switches. Wiley, Weinheim, p 219
58. Castro R, Nixon KR, Evanseck JD, Kaifer AE (1996) J Am Chem Soc 61:7298
59. Wudl F, Smith GM, Hufnagel EJ (1970) Chem Commun 1453
60. Philp D, Slawin AMZ, Spencer N, Stoddart JF (1991) Chem Commun 1584
61. Ballardini R, Balzani V, Dehaen W, Dell'erba AE, Raymo FM, Stoddart JF, Venturi M (2000) Eur J Org Chem 591
62. Collier CP, Mattersteig G, Wong EW, Luo Y, Beverly K, Sampaio J, Raymo FM, Stoddart JF, Heath JR (2000) Science 289:1172
63. Wong EW, Collier CP, Behloradsky M, Raymo FM, Stoddart JF, Heath JR (2000) J Am Chem Soc 122:5831
64. Collier CP, Jeppesen JO, Luo Y, Perkins J, Wong EW, Heath JR, Stoddart JF (2001) J Am Chem Soc 123:12632
65. Diehl MR, Steuerman DW, Tseng HR, Vignon SA, Star A, Celestre PC, Stoddart JF, Heath JR (2003) ChemPhysChem 3:519
66. Choi JW, Steuerman DW, Nygaard S, Flood AH, Braunschweig AB, Moonen NNP, Laursen BW, Luo Y, DeIonno E, Peter AJ, Jeppesen JO, Stoddart JF, Heath JR (2005) Chem Eur J, accepted
67. Asakawa M, Dehaen W, L'abbe G, Menzer S, Nouwen J, Raymo FM, Stoddart JF, Williams DJ (1996) J Org Chem 61:9591
68. Anelli PL, Spencer N, Stoddart JF (1991) J Am Chem Soc 113:5131
69. Fyfe MCT, Stoddart JF (1997) Acc Chem Res 30:393
70. Bissell RA, Córdova E, Kaifer AE, Stoddart JF (1994) Nature 369:133
71. Nielsen MB, Jeppesen JO, Lau J, Lomholt C, Damgaard D, Jacobsen JP, Becher J, Stoddart JF (2001) Potential vs. Ag/AgCl in MeCN. K_a determined by UV/Vis dilution studies in MeCN at 22 °C. J Org Chem 66:3559
72. Córdova E, Bissell RA, Kaifer AE (1995) Potential vs. Ag/AgCl in MeCN. J Org Chem 60:1033
73. Balzani V, Credi A, Mattersteig G, Matthews OA, Raymo FM, Stoddart JF, Venturi M, White AJP, Williams DJ (2000) J Org Chem 65:1924
74. Anelli PL, Ashton PR, Ballardini R, Balzani V, Delgado M, Gandolfi MT, Goodnow TT, Kaifer AE, Philp D, Pietraszkiewicz M, Prodi L, Reddington MV, Slawin AMZ, Spencer N, Stoddart JF, Vicent C, Williams DJ (1992) J Am Chem Soc 114:193
75. Córdova E, Bissell RA, Spencer N, Ashton PR, Stoddart JF, Kaifer AE (1993) J Org Chem 58:6550
76. Devonport W, Blower MA, Bryce MR, Goldenberg LM (1997) K_a determined by UV/Vis dilution studies in MeCN at 21 °C. J Org Chem 62:885

77. Mirzoian A, Kaifer AE (1995) J Org Chem 60:8093
78. Smitzrud DB, Diederich F (1990) J Am Chem Soc 112:339
79. Kosower EM (1958) J Am Chem Soc 80:3253
80. Reichardt C (1988) In: Solvents and Solvent Effects in Organic Chemistry, 2nd ed. Wiley, Weinheim, chap 7, p 339
81. Kang S, Vignon SA, Tseng H-R, Stoddart JF (2004) Chem Eur J 10:2555
82. Tseng HR, Vignon SA, Stoddart JF (2003) Angew Chem Int Ed 42:1491
83. Tseng HR, Vignon SA, Celestre PC, Perkins J, Jeppesen JO, Di Fabio A, Ballardini R, Gandolfi MT, Venturi M, Balzani V, Stoddart JF (2004) Chem Eur J 10:155
84. Tseng HR, Wu D, Fang NX, Zhang X, Stoddart JF (2004) ChemPhysChem 5:111
85. Flood AH, Stoddart JF, Steuerman DW, Heath JR (2004) Science 306:2055
86. Steuerman DW, Tseng HR, Peters AJ, Flood AH, Jeppesen JO, Nielsen KA, Stoddart JF, Heath JR (2004) Angew Chem Int Ed 43:6486
87. Flood AH, Peters AJ, Vignon SA, Steuerman DW, Tseng HR, Kang S, Heath JR, Stoddart JF (2004) Chem Eur J 10:6558
88. Jang SS, Jang YH, Kim YH, Goddard III WA, Flood AH, Laursen BW, Tseng HR, Stoddart JF, Jeppesen JO, Choi JW, Steuerman DW, Delonno E, Heath JR (2005) J Am Chem Soc 127:1563
89. Jeppesen JO, Perkins J, Becher J, Stoddart JF (2001) Angew Chem Int Ed 40:1216
90. Jeppesen JO, Nielsen KA, Perkins J, Vignon SA, Di Fabio A, Ballardini R, Gandolfi MT, Venturi M, Balzani V, Becher J, Stoddart JF (2003) Chem Eur J 9:2982
91. Jeppesen JO, Vignon SA, Stoddart JF (2003) Chem Eur J 9:4611
92. Jeppesen JO, Nygaard S, Vignon SA, Stoddart JF (2005) Eur J Org Chem 196
93. Laursen BW, Nygaard S, Jeppesen JO, Stoddart JF (2004) Org Lett 6:4167
94. Anelli PL, Asakawa M, Ashton PR, Bissell RA, Clavier G, Górski R, Kaifer AE, Langford SJ, Mattersteig G, Menzer S, Philp D, Slawin AMZ, Spencer N, Stoddart JF, Tolley MS, Williams DJ (1997) Chem Eur J 3:1113
95. Yamamoto T, Tseng HR, Stoddart JF, Balzani V, Credi A, Marchioni P, Venturi M (2003) There are indications for a folded conformation of the rotaxane in which there are $\pi-\pi$ interactions between the TTF and the outer π-surface of the tetracation, see: Collect Czech Chem Commun 68:1488
96. Ashton PR, Blower M, Philp D, Spencer N, Stoddart JF, Tolley MS, Ballardini R, Ciano M, Balzani V, Gandolfi MT, Prodi L, McLean CH (1993) New J Chem 17:689
97. As a generalization, the absolute values of $\Delta(\Delta H_a)$ and $\Delta(\Delta S_a)$ are taken. This scheme would not be valid in the extreme case in which $\Delta(\Delta H_a)$ and $\Delta(\Delta S_a)$ are both large and opposite in sign. In that case, always very good population ratios will be found. The design of such a system would be a major challenge
98. Kramers H (1940) Physica 7:284
99. Grote RF, Hynes JT (1980) J Chem Phys 73:2715
100. Grote RF, Hynes JT (1981) J Chem Phys 74:4465
101. Sumi H, Marcus RA (1986) J Chem Phys 84:4894
102. Sumi H (1995) J Mol Liq 65/66:65
103. Ansari A, Jones CM, Henry ER, Hofrichter J, Eaton WA (1992) Science 256:1796
104. Jacob M, Greeves M, Holtermann G, Schmid FA (1999) Nat Struct Biol 6:923

Hydrogen Bond-Assembled Synthetic Molecular Motors and Machines

Euan R. Kay · David A. Leigh (✉)

School of Chemistry, The University of Edinburgh, West Mains Road, Edinburgh EH9 3JJ, UK
David.Leigh@ed.ac.uk

1	**Design Principles for Molecular-Level Motors and Machines**	134
1.1	Molecular-Level Machines and the Language Used to Describe Them	135
1.2	The Effects of Scale	136
1.3	Lessons to Learn from Biological Motors and Machines	138
2	**Mechanically Interlocked Molecular Architectures**	139
2.1	Basic Features of Catenanes and Rotaxanes	139
2.2	Amide-Based Catenanes and Rotaxanes	140
3	**Controlling Motion in Amide-Based Catenanes and Rotaxanes**	143
3.1	Inherent Dynamics: Pirouetting in Benzylic Amide Catenanes	143
3.2	Inherent Dynamics: Ring Pirouetting in Benzylic Amide Rotaxanes	144
3.3	Inherent Dynamics: Shuttling in Benzylic Amide Rotaxanes	145
3.3.1	Shuttling in Degenerate, Two Binding Site, Peptide-Based Molecular Shuttles	146
3.3.2	A Physical Model of Degenerate, Two Binding Site, Molecular Shuttles	147
3.4	Translational Molecular Switches: Stimuli-Responsive Molecular Shuttles	149
3.4.1	A Physical Model of Two Binding Site, Stimuli-Responsive Molecular Shuttles	149
3.4.2	Adding and Removing Protons to Induce Net Positional Change	151
3.4.3	Adding and Removing Electrons to Induce Net Positional Change	152
3.4.4	Adding and Removing Covalent Bonds to Induce Net Positional Change	154
3.4.5	Changing Configuration to Induce Net Positional Change	155
3.4.6	Entropy-Driven Net Positional Change	155
3.5	Controlling Rotational Motion: Ring Pirouetting in Rotaxanes	157
3.6	Controlling Rotational Motion in Catenanes	159
3.6.1	Two-Way and Three-Way Catenane Positional Switches	159
3.6.2	Directional Circumrotation: A [3] Catenane Rotary Motor	161
3.6.3	Selective Rotation in Either Direction: A [2] Catenane Reversible Rotary Motor	163
4	**Property Effects Using Amide-Based Synthetic Molecular Machines**	167
4.1	Switching On and Off Induced Circular Dichroism with a Molecular Shuttle	167
4.2	Switching On and Off Fluorescence with a Molecular Shuttle	168
4.3	Rotaxane-Based Photoresponsive Surfaces and Macroscopic Transport by Molecular Machines	171

| 5 | Conclusions | 174 |
| References | | 174 |

Abstract Nature uses molecular motors and machines in virtually every significant biological process but learning how to design and assemble simpler artificial structures that function through controlled molecular-level motion is a major challenge for contemporary physical science. In this review we discuss some of the principles behind synthetic molecular motors and machines and examine a class of molecular architectures, benzylic amide catenanes and rotaxanes, that are proving promising in this area. The movement of the components in these systems can be controlled by light, electrons, chemical reactions, pH, temperature and the nature of the environment leading to both simple switches (molecular shuttles) and more complex molecular motors. They operate through biasing random thermal motion and can be understood through an appreciation of physical fluxional transport mechanisms. Remarkably simple examples of stimuli-responsive molecular shuttles can be interfaced with—and even perform physical tasks in—the macroscopic world.

Keywords Catenane · Molecular machine · Molecular motor · Molecular shuttle · Rotaxane

Abbreviations
a.c.   alternating current
ICD   induced circular dichroism
NMR   nuclear magnetic resonance
NTs N-tosyl
SPT-SIR spin polarization transfer by selective inversion recovery
VT   variable temperature

1
Design Principles for Molecular-Level Motors and Machines

The widespread use of molecular-level motion in key natural processes suggests that great rewards could come from bridging the gap between the present generation of synthetic molecular systems—which by and large rely upon electronic and chemical effects to carry out their functions—and the machines of the macroscopic world, which utilize the synchronized movements of smaller parts to perform particular tasks. In recent years it has proved relatively straightforward to design synthetic molecular systems where positional changes of submolecular components occur by moving energetically downhill, but what are the structural features necessary for molecules to use directional displacements to do work? How can we make a synthetic molecular machine that pumps ions against a gradient, say, or moves itself energetically uphill along a track? Artificial compounds that can do such things have yet to be realized; the field of synthetic molecular machines is still in its infancy and only the most basic systems—mechanical switches and slightly more sophisticated, but still rudimentary, molecular rotary motors—have been made thus far [1–6]. Here we outline some early successes in taming molecular-level movement using a particular type of molecular architecture, benzylic amide catenanes and rotaxanes.

1.1
Molecular-Level Machines and the Language Used to Describe Them

Language—especially scientific language—has to be suitably defined and correctly used in order to accurately convey concepts in a field. Nowhere is the need for accurate scientific language more apparent than in the discussion of the ideas and mechanisms by which nanoscale machines could—and do—operate. Much of the terminology used to describe molecular-level machines comes from the systems observed by physicists and biologists, but unfortunately their findings and descriptions have sometimes been misunderstood or unappreciated by chemists. Perhaps inevitably in a newly emerging field, there is not even a clear consensus as to what constitutes a molecular machine and what differentiates them from other molecular devices [1–6]. Initially, the categorization of molecules as machines by chemists was purely iconic—the structures "looked" like pieces of machinery—or they were so-called because they carried out a function that in the macroscopic world would require a machine to perform it. Many of the chemical systems first likened to pistons and other machines were simply host-guest complexes in which the binding could be switched "on" or "off" by external stimuli such as light. Whilst these early studies were the key to popularizing the field, with hindsight a consideration of the effects of scale tells us that supramolecular decomplexation events have little in common with the motion of a piston (the analogy is better within a rotaxane architecture because the components are still kinetically associated after decomplexation but the implication of imparting momentum is still unfortunate) and that a photosensitizer is not phenomenologically related to a "light-fueled motor". In fact, it is probably most useful to differentiate "device" and "machine" on the basis that the etymology and meaning of machine implies mechanical movement—i.e. a net nuclear displacement in the molecular world—which causes something useful to happen. This leads to the definition that "molecular machines" are a subset of "molecular devices" (functional molecular systems) in which some stimulus triggers the controlled, large amplitude or directional mechanical motion of one component relative to another (or of a substrate relative to the machine) which results in a net task being performed.

When developing the terminology necessary to describe molecular behavior or systems scientifically, the standard dictionary definitions meant for everyday use are not always appropriate in regimes that the definitions were never intended to cover. The difference between a "molecular motor" and a "molecular switch", for example, is crucial because they are different phenomenological descriptors, not iconic classifications of molecular shapes: In physical terms a "switch" influences a system as a function of state; a "motor" influences a system as a function of its trajectory. Returning a switch to its original position undoes its effect on an external system, whereas when a motor returns to its original position through a different pathway to the one it left by (say, a 360° directional rotation in the case of a rotary motor), it does not. This difference is profound and the terms really should not be used interchangeably as sometimes happens in the chemistry (but not physics or biology) literature. Switches cannot use chemical energy to progressively drive a system away from equilibrium whereas a motor can.

The examples of amide-based catenanes and rotaxanes in this Chapter are chosen to help demonstrate some of the requirements for mechanical task performance at the molecular level.

1.2
The Effects of Scale

The path towards synthetic molecular machines starts nearly two centuries ago with the observation of effects that pointed to the random motion inherently experienced by all molecular-scale structures. In 1827, the Scottish botanist Robert Brown noted through his microscope the incessant, haphazard motion of tiny particles within translucent pollen grains suspended in water. An explanation of the phenomenon—now known as Brownian motion or movement—was provided by Einstein in one of his three extraordinary papers of 1905 and was proved experimentally by Perrin over the next decade. Ever since, scientists have been fascinated by the stochastic nature of molecular-level motion and its implications. The random thermal fluctuations experienced by molecules dominate mechanical behavior in the molecular world. Even the most efficient nanoscale machines are swamped by its effect. A typical motor protein consumes ATP fuel at a rate of 100–1000 molecules every second, corresponding to a maximum possible power output in the region 10^{-16} to 10^{-17} W per molecule. When compared with the random environmental buffeting of $\sim 10^{-8}$ W experienced by molecules in solution at room temperature, it seems remarkable that *any* form of controlled motion is possible [7]. When designing molecular machines it is important to remember that the presence of Brownian motion is a consequence of scale, not of the nature of the surroundings. It cannot be avoided by putting a molecular-level structure in a near-vacuum for example. Although there would be few random collisions to set such a Brownian particle in motion, equally there would be little viscosity to slow it down. These effects always cancel each other out and as long as a temperature for an object can be defined, it will undergo Brownian motion appropriate to that temperature (which determines the kinetic energy of the particle). In the absence of *any* other molecules, heat would still be transmitted from the hot walls of the container to the particle by electromagnetic radiation, the random emission and absorption of the photons producing the Brownian motion. In fact, even temperature is not a very effective modulator of the background Brownian motion since the velocity of the particles depends on the square root of the temperature. So to reduce Brownian motion to 10% of the amount present at room temperature, one would have to drop the temperature from 300 K to 3 K [7, 8].

It seems sensible, therefore, to try to utilize Brownian motion when designing molecular machines rather than make structures that have to fight against it. In exciting developments over the past decade, theoretical physics has explained how random, *directionless* fluctuations can cause the *directional* transport of particles [7, 9–14] successfully accounting for the general principles behind biological motors [15–18]. An appreciation of the physics involved is crucial for the design of artificial structures which perform mechanical operations at the molecular level and transpose those effects to the macroscopic world. Nature provides examples of successful designs in the form of ion pumps, motor proteins, photoactive proteins, retinal, and many other systems [19]. Indeed, biological structures have already been incorporated into semi-synthetic biomaterials which can carry out "unnatural" mechanical tasks [20, 21].

The constant presence of Brownian motion is not the only distinction between motion at the molecular level and in the macroscopic world. The physics which govern mechanical dynamic processes in the two regimes are completely different, therefore requiring fundamentally different mechanisms for controlled transport or propulsion. In the macroscopic world, the equations of motion are governed by inertial terms (dependent on mass). Viscous forces (dependent on surface areas) dampen motion by converting kinetic energy into heat, and objects do not move until provided with specific energy to do so. In a macroscopic machine this is often provided through a directional force when work is done to move mechanical components in a particular way. As objects become less massive and smaller in dimension, inertial terms decrease in importance and viscosity begins to dominate. A parameter which quantifies this effect is Reynolds number (R)—essentially the ratio of inertial to viscous forces—given by Eq. 1 for a particle of length dimension a, moving at velocity v, in a medium with viscosity η and density ρ [22].

$$R = \frac{av\rho}{\eta} \quad (1)$$

Size affects modes of motion long before we reach the nanoscale. Even at the mesoscopic level of bacteria (length dimensions $\sim 10^{-5}$ m), viscous forces dominate. At the molecular level, Reynolds number is extremely low (except at low pressures in the gas phase) and the result is that molecules, or their components, cannot be given a one-off "push" in the macroscopic sense—momentum is irrelevant. The motion of a molecular-level object is determined entirely by the forces acting on it at that particular instant—whether they be externally applied forces, viscosity, or random thermal perturbations and Brownian motion. In more general terms, this analysis leads to a central tenet: while the macroscopic machines we encounter in everyday life may provide the inspiration for what we might like molecular machines to achieve, drawing too close an analogy for their modes of operation is a poor design strategy. The "rules of the game" at different length scales are simply too different. Two basic principles must be followed for any molecular device to be able to carry out a mechanical function: First, the movement of the kinetically-associated molecules or their components must be controlled by employing interactions which restrict the natural tendency for three-dimensional random motion and somehow bias, rectify or direct the motion along the required vectors. Secondly, the Second Law of Thermodynamics tells us that no machine can continually operate solely using energy drawn from the thermal bath, so an external input of energy is required to perform a mechanical task with any synthetic molecular machine.

1.3
Lessons to Learn from Biological Motors and Machines

That the miniaturization of mechanical motors and machines to the molecular level is feasible is demonstrated by the fact that they are already all around us. Nature has developed an exquisite molecular nanotechnology that it employs to astonishing effect in virtually every significant biological process. Appreciating how Nature has overcome the issues of scale, environment, Brownian motion, and viscosity is extremely useful for learning how to design synthetic molecular systems that have to do the same.

There are many important differences between biological molecular machines and the man-made machines of the macroscopic world: Biological machines are soft, not rigid; they work at ambient temperatures (heat is dissipated almost instantaneously at small length scales so one cannot exploit temperature gradients); biological motors utilize chemical energy, often in the form of ATP hydrolysis or chemical gradients; they work in solution or at surfaces and operate under conditions of intrinsically high viscosity; they rely on and utilize—rather than oppose—Brownian motion; biomolecular machines need not use chemical energy to initiate movement—their components are constantly in motion—rather, they must control the directionality of that movement; the molecular machine and the substrate(s) it is acting upon are kinetically associated during the operation of the machine; biological machines are made by a combination of multiple parallel synthesis and self-assembly; their operation is governed by non-covalent interactions; and, finally, they utilize architectures (e.g. tracks) which restrict most of the degrees of freedom of the machine components and/or the substrate(s) it is acting upon. It is these last points, in particular, that have encouraged the study of catenanes and rotaxanes as potential structures for synthetic molecular motors and machines.

2
Mechanically Interlocked Molecular Architectures

2.1
Basic Features of Catenanes and Rotaxanes

Catenanes are chemical structures in which two or more macrocycles are interlocked, while in rotaxanes, one or more macrocycles are mechanically prevented from de-threading from linear chains by bulky "stoppers" (Fig. 1) [23–28]. Even though the components are not covalently connected, catenanes and rotaxanes are molecules—not supramolecular complexes—as covalent bonds must be broken in order to separate the constituent parts. In these kinetically associated species [29], the mechanical bond imparts a restriction in the degrees of freedom for relative movement of the components while often permitting extraordinarily large amplitude motion in the allowed vectors. This is in many ways analogous to the restriction of movement imposed on biological motors by a structural track [30–32].

Large-amplitude submolecular motions in catenanes and rotaxanes can be divided into two classes (Fig. 1): pirouetting of the macrocycle around the thread (rotaxanes) or another ring (catenanes); and translation of the macrocycle along the thread (rotaxanes) or around another ring (catenanes). By analogy to conformational changes within classical molecules, the relative movements between interlocked species are termed "co-conformational" changes [33]. Motions in these systems are most easily described by taking one component as a frame of reference, around or along which the other parts

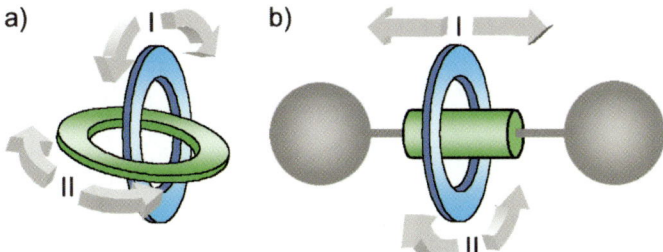

Fig. 1 Schematic representation of **a** a [2]catenane and **b** a [2]rotaxane. *Arrows* show possible large amplitude modes of movement of one component relative to another

move. Often the choice of this frame of reference is based on the relative size or complexity of the components, although this distinction can become blurred.

For a long time synthetic approaches to such structures relied on statistical or relatively inefficient covalently-directed approaches [34–37]. The development of supramolecular chemistry, however, allowed chemists to apply their increasing appreciation of non-covalent interactions to synthesis, resulting in numerous, often extremely elegant, template methods to catenanes and rotaxanes [38–51]. In such syntheses, non-covalent binding interactions between the components often "live-on" in the interlocked products. These interactions restrict the relative motions of components further to that defined by their architecture and can ultimately be manipulated to effect positional displacements. Much attention has been given to both intrinsic submolecular motions within these structures and the control of the relative positioning of the interlocked components through manipulation of recognition events.

2.2
Amide-Based Catenanes and Rotaxanes

Amide-based catenanes, in which hydrogen bonds template interlocking of the two rings, were first reported in 1992 by Hunter (**1**, Fig. 2) [52]. The steric bulk of the cyclohexyl groups prevents circumrotation of the macrocyclic rings in **1**, a feature which allowed Hunter to suggest the mechanically interlocked archi-

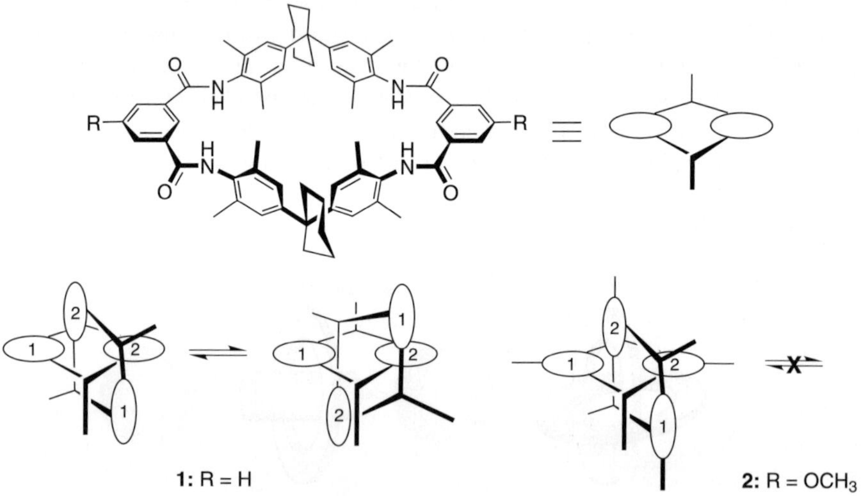

Fig. 2 Hydrogen bond-assembled [2]catenanes **1** and **2** reported by Hunter and Vögtle respectively, showing the limited component rotation possible in **1** but not in **2**, which is essentially completely locked in conformational terms

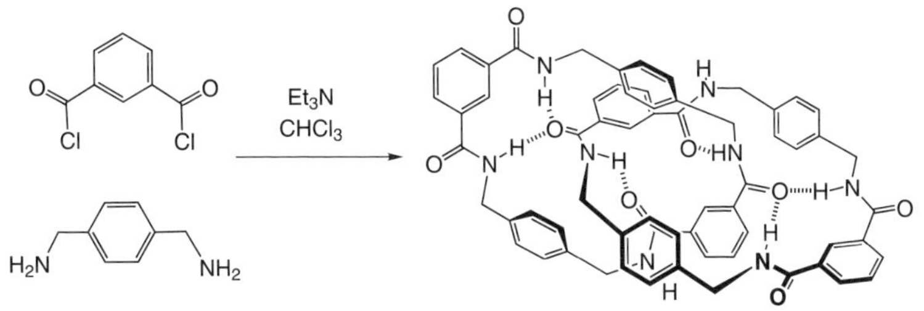

Scheme 1 The original hydrogen bond-assembled benzylic amide [2]catenane **3**, made in one step (20% yield) from isophthaloyl dichloride and *p*-xylylene diamine

tecture through a quite brilliant analysis of the complex 1D and 2D ^{1}H NMR data. Hunter's discovery prompted the publication of similar compounds by Vögtle (**2**) [53]. In the Hunter catenane (**1**), a 90° rotational motion which exchanges the non-equivalent isophthaloyl rings (labeled "1" and "2" in Fig. 2) is possible, but even this is blocked on substitution at the 5-position of the isophthaloyl ring (e.g. in **2**). Despite the accumulation of a large body of work over the subsequent decade, the fundamental problem that these systems exist in a rigidly locked co-conformation has never been overcome.

Fortunately, the benzylic amide catenane **3** (Scheme 1) discovered by chance in 1995 exhibits far more versatile dynamics than the Hunter–Vögtle system [54]. Catenane **3** has the simplest possible ^{1}H NMR spectrum, with only the six different constitutional types of protons apparent in [D$_6$]DMSO, the same number seen for the parent macrocycle [55]. Despite the smaller macrocycle cavity size of **3** compared to **1** and **2**, both the isophthaloyl and *p*-xylylene components are able to rotate through the cavity of the other macrocycle and the circumvolution process is fast on the NMR timescale in polar solvents at room temperature. The catenane formation process tolerates a number of aromatic 1,3-dicarbonyl and benzylic amide precursors so that a diverse range of analogues could be prepared (e.g. **4**, Fig. 3) and the effect of structure on the dynamic processes determined [56].

The hydrogen bonding patterns in the X-ray crystal structures of the benzylic amide catenanes suggested that modification of the synthetic procedure might allow rotaxanes to be synthesized. Sure enough, if the catenane-forming reaction was carried out in the presence of a suitable stoppered benzylic 1,3-diamide thread (e.g. **5**) rotaxanes such as **6** were also formed (Scheme 2) [55].

The rotaxane architecture brings about significant changes in the chemical and physical properties of the components; for example, the rotaxane **6** is 10^5 times more soluble in non-polar organic solvents than the parent macrocycle [55].

The wealth of structural data obtained on the benzylic amide catenane and rotaxane system showed that the mechanism for the formation of the

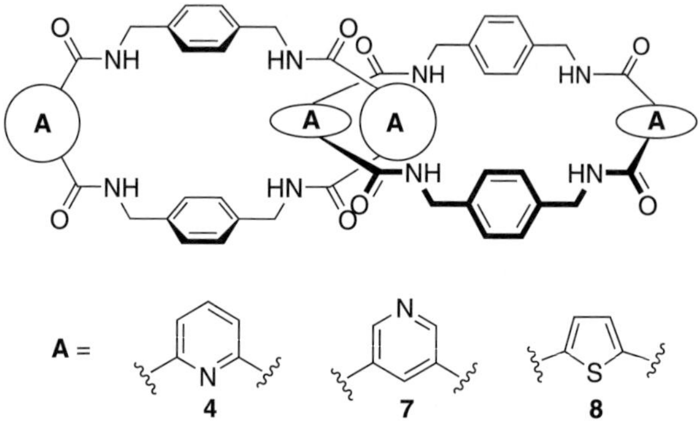

Fig. 3 Benzylic amide catenanes **4**, **7** and **8** with differing aromatic-1,3-dicarbonyl units and differing dynamic properties

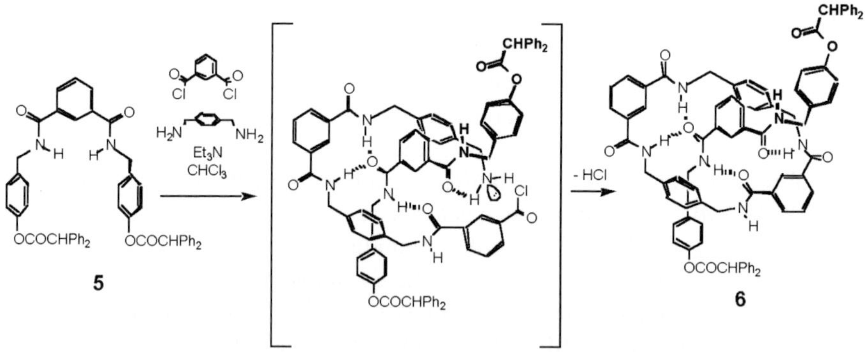

Scheme 2 Hydrogen bond-directed benzylic amide rotaxane formation

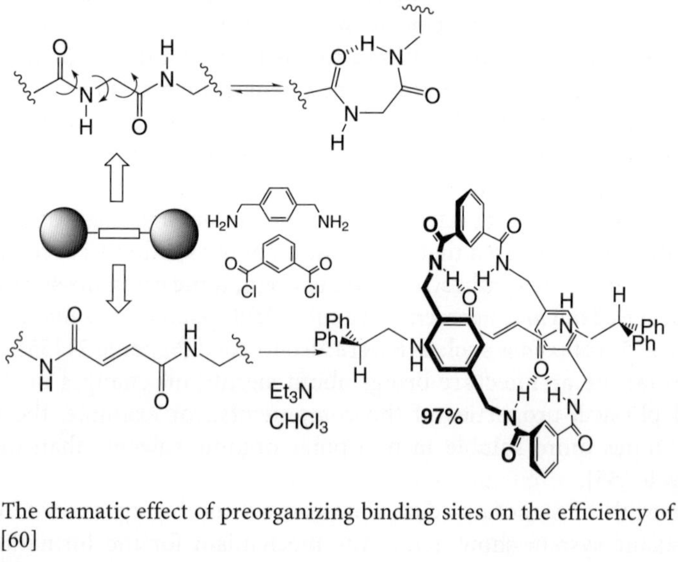

Scheme 3 The dramatic effect of preorganizing binding sites on the efficiency of rotaxane synthesis [60]

interlocked products was primarily the templated assembly of the macrocycle around two amide bonds [57]. Divergent hydrogen bonding sites in a similar spatial arrangement occur in adjacent amino acid residues in peptide chains and, indeed, it was found that simple peptides are also able to template the cyclization of benzylic amide macrocycles to give peptide rotaxanes [58, 59]. However, the inherent flexibility of the peptide backbone is not optimal for rotaxane formation. Replacing the peptide motif with a fumaramide unit maintains the required hydrogen bonding groups but with a rigidified backbone. This results in a remarkable 97% yield for a rotaxane, from a five component assembly process in just two steps from commercial starting materials (Scheme 3) [60].

3
Controlling Motion in Amide-Based Catenanes and Rotaxanes

3.1
Inherent Dynamics: Pirouetting in Benzylic Amide Catenanes

Any alkyl substitution on the isophthaloyl ring (at either the 5- or 4-positions) prevents complete circumvolution of the macrocycles in benzylic amide catenanes, thereby destroying the plane of symmetry bisecting the isophthaloyl units in each ring. A series of coalescence temperature measurements and spin polarization transfer by selective inversion recovery (SPT-SIR) NMR experiments [56, 61], revealed that the pyridine-2,6-dicarbonyl derivative (**4**) exhibits slow circumvolution on the NMR timescale. The thiophene derivative (**8**) was shown to rotate the fastest—in fact 3.2 million times faster than **4** in $C_2D_2Cl_4$ at room temperature.

This remarkable variation in dynamic behavior resulting from relatively simple structural changes can be allied to more subtle effects on rate using the solvent composition. Hydrogen bond-disrupting solvents such as CD_3OD and, to an even greater degree, $[D_6]DMSO$ increase the rate of circumvolution by competing for the hydrogen bonding groups in the macrocycle. This has the effect of weakening the ground-state interactions and stabilizing the intermediate co-conformations during circumvolution, therefore lowering the energy barriers for the process. Variation of solvent composition allows fine-tuning of the circumvolution rate following selection of the gross "ball-park" value through structure choice [56, 61].

Both these solvent and structural effects on rate suggest that a key process in the circumvolution mechanism is rupture and formation of intercomponent hydrogen bonds. An isolated molecule molecular mechanics approach has been used to simulate the dynamics in these catenane systems, allowing simulation of the molecular shape at the transition states [61, 62]. Together with a full low-dimensional quantum-mechanical description of circumvolution [63], these studies provide the necessary theoretical and practical understanding of the intrinsic motions in a catenane system important for the development of molecular devices in which the frequency of motions may be a key characteristic.

3.2
Inherent Dynamics: Ring Pirouetting in Benzylic Amide Rotaxanes

Pirouetting is the random Brownian rotation of the macrocycle about the axis defined by the thread in a rotaxane. While this motion is often clearly occurring, it can be difficult to study in detail because of the symmetry of the components. Benzylic amide macrocycle-based rotaxanes possess a useful characteristic in this respect. In many examples, the benzylic amide macrocycle adopts a chair-like conformation meaning that for each pair of benzylic protons (H_E, Fig. 4), one is in an equatorial environment, while the other is axial [58]. For a macrocycle on a symmetric thread, two ^{1}H NMR signals would therefore be expected for the 8 benzylic protons in the molecule. Rotation of the macrocycle around the axis shown by 180° must, however, result in a chair-chair flip so as to maintain the hydrogen bonding network between the macrocyclic amide protons and carbonyl oxygens on the thread. This co-conformational change therefore interconverts the axial and equatorial sets of protons twice during a full 360° revolution. It is possible to study chemical exchange processes such as this by variable temperature (VT) NMR techniques, including the coalescence method [64], or spin polarization transfer by selective inversion recovery (SPT-SIR) [65].

For example, the room temperature ^{1}H NMR spectrum in CDCl$_3$ of the glycylglycine-based [2]rotaxane **9** displays the fewest possible signals for the macrocycle protons, indicating rapid pirouetting (820 s^{-1}) of the ring at ambient temperature [58]. The pyridyl-2,6-dicarbonyl-based macrocycle in **10**

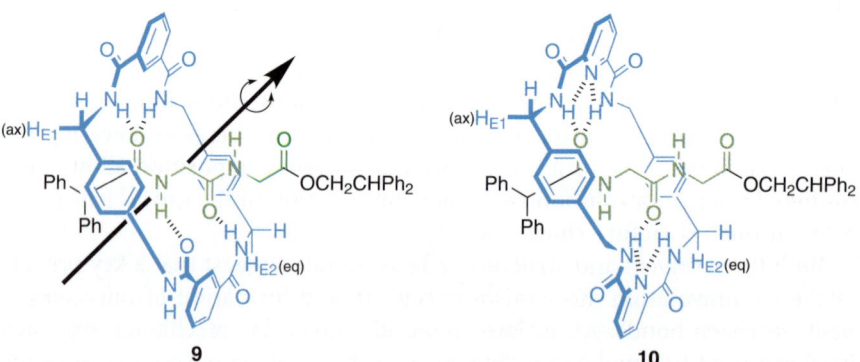

Fig. 4 Peptidorotaxanes **9** and **10**. The *arrow* on **9** indicates the axis about which macrocycle pirouetting occurs

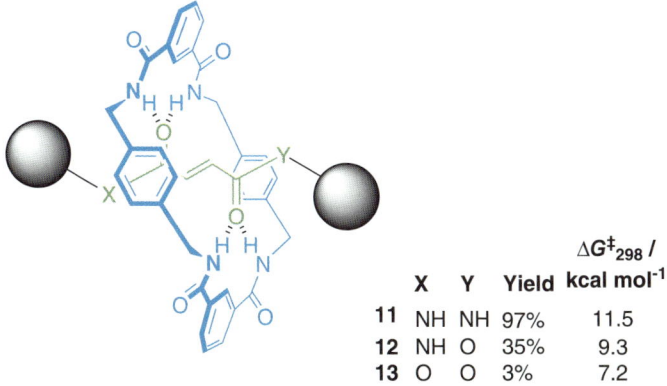

Fig. 5 Variation of yield in the rotaxane forming reactions for a series of simple fumaric acid based templates is correlated with the strength of intercomponent interactions in the products, as evidenced by $\Delta G^{\ddagger}_{298}$, the energy barrier to pirouetting in CD_2Cl_2

possesses a different, overall slightly stronger, hydrogen bonding network between the components, producing a concomitant reduction in pirouetting rate to $100\,s^{-1}$ in $CDCl_3$ at rt.

As the rate of pirouetting is directly related to the strength with which the macrocycle is held to the thread, this motion can be a useful probe to evaluate the effect of structural changes on the strength of intercomponent interactions in rotaxanes. For the series of fumaric acid based [2]rotaxanes shown in Fig. 5, yields of the rotaxane forming process (a five-component clipping reaction in which formation of the benzylic amide macrocycle is templated by the fumaric acid residue) increase in the order 13 < 12 < 11 [60]. As expected, the rates of macrocycle pirouetting in the products were shown to follow exactly the opposite trend (Fig. 5), confirming that the efficiency of the templated synthesis is directly related to the hydrogen bonding ability of the thread templates.

3.3
Inherent Dynamics: Shuttling in Benzylic Amide Rotaxanes

Shuttling is the movement of a macrocycle back and forth along the thread component of a rotaxane. This motion takes the form of a Brownian motion-powered random walk, constrained to one dimension by the thread and to translational displacement boundaries by the bulky stoppers. By virtue of the template methods employed in their synthesis, the threads of amide-based rotaxanes consist of one or more recognition elements, or "stations" for the macrocycle(s); shuttling therefore becomes the movement on, off, or between such stations and just as for pirouetting motions, the dynamics depend on the strength of the intercomponent interactions.

3.3.1
Shuttling in Degenerate, Two Binding Site, Peptide-Based Molecular Shuttles

Following the observation that hydrogen bond-disrupting solvents such as [D$_6$]DMSO destroy the interactions between the benzylic amide macrocycle and a glycylglycine thread in [2]rotaxane **9** (Fig. 4) [58], a series of peptide-based molecular shuttles were reported in which two glycylglycine stations for the benzylic amide macrocycle are separated by aliphatic linkers [59]. In **14–16** (Fig. 6), in CDCl$_3$, the macrocycle shuttles between the two degenerate peptide stations rapidly at room temperature, evidenced by the single set of signals for the two peptide stations in the ^{1}H NMR spectrum which is resolved into two sets (for the occupied and unoccupied stations) on cooling the sample and freezing out the motion on the NMR timescale.

Just as for the pirouetting motions, the shuttling mechanism requires at least partial rupture of the intercomponent interactions at one station, before formation of new interactions at the new station. The thermodynamics of the degenerate process in each of **14–16** therefore involves passage over some activation barrier from one energy well to another identical minimum (Fig. 7). Of course, if the kinetic barrier is significantly larger than the thermal energy, shuttling ceases to occur. This can be achieved in **17** by introduction of a bulky *N*-tosyl (NTs) moiety (Fig. 6). Shuttling is restored on removal of this bulky barrier. Just as for the analogous benzylic amide catenanes, solvent composition has a profound effect on the shuttling rate in **14–16**— as little as 5% CD$_3$OD in halogenated solutions of peptide-based rotaxanes provides rate increases in excess of two orders of magnitude. The hydrogen bond-disrupting methanol weakens the intercomponent interactions, effectively loosening the macrocycle from its station (i.e. reducing $\Delta G^\ddagger$) with a concomitant increase in the shuttling rate [59].

This increase in rate does not, however, continue indefinitely with increasing solvent hydrogen bond basicity. An additional feature of these rotaxanes is

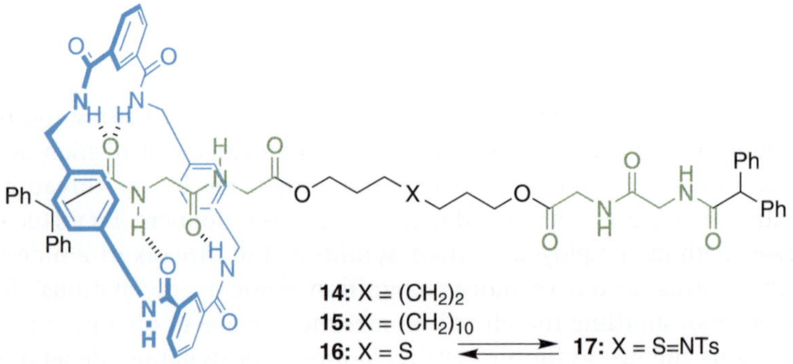

14: X = (CH$_2$)$_2$
15: X = (CH$_2$)$_{10}$
16: X = S   ⇌   **17:** X = S=NTs

Fig. 6 Shuttling in peptide-based degenerate station molecular shuttles **14–17**

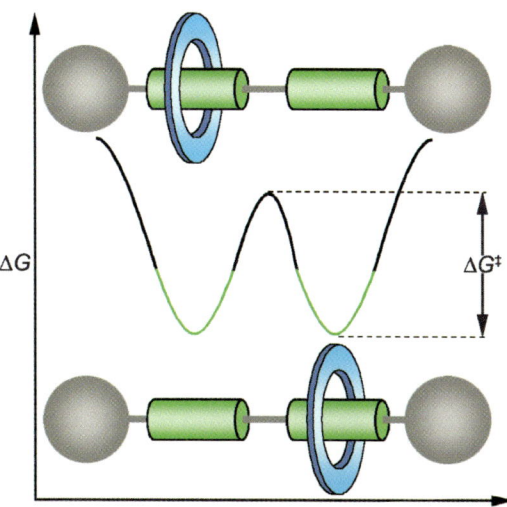

Fig. 7 Idealized free energy profile for movement between two identical stations in a degenerate molecular shuttle. The height of the barrier $\Delta G^{\ddagger}$ contains two components: the energy required to break the non-covalent interactions holding it to the station; and a distance-dependent diffusional component

that a major change in solvent polarity (changing from halogenated solvents to [D_6]DMSO) stops the macrocycle shuttling between the peptide stations and causes it to preferentially reside over the hydrophobic thread instead, hiding the thread from the unfavorably polar environment and allowing maximum hydrogen bonding between the peptide stations and solvent [59].

3.3.2
A Physical Model of Degenerate, Two Binding Site, Molecular Shuttles

An interesting structural effect is observed on increasing the length of the spacer in degenerate shuttles. The rate of shuttling in [2]rotaxane **14** was compared to that of **15** (Fig. 6). Although ostensibly not involved in any interactions with the macrocycle, extension of the alkyl chain results in an experimentally measured reduction in the rate of shuttling which corresponds to an increase in activation energy for the process of 1.2 kcal mol^{-1} at rt—an effect solely of the increased distance the macrocycle must travel [59].

These effects are most easily understood by considering the macrocycle as a particle moving along a one-dimensional potential energy (rather than free energy) surface (Fig. 8). At any point on the potential energy surface, the gradient of the line gives the force exerted on the macrocycle by the thread. When the macrocycle is in the vicinity of a station, hydrogen bonds and/or other attractive non-covalent interactions exert large forces (typically varying as a high power of the inverse distance, r^{-n}) on the macrocycle, opposing its mo-

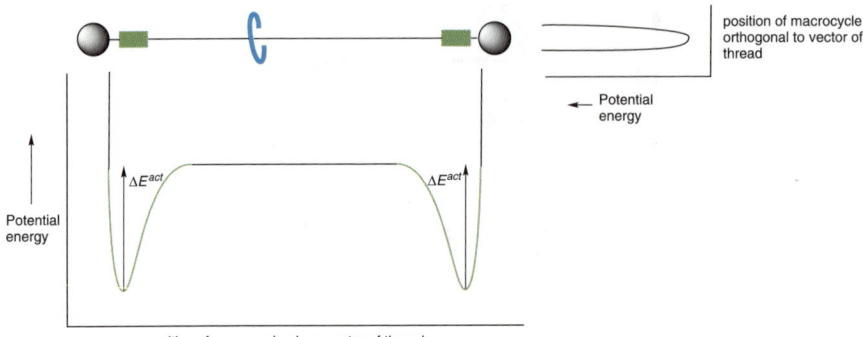

Fig. 8 Idealized potential energy of the macrocycle in a two-station degenerate molecular shuttle. The potential energy surface shows the effect of the interaction between macrocycle and thread on the energy of the macrocycle (ignoring any complicating factors such as folding). Chemical potential energies (ΔE's) generally follow similar trends to free energies (ΔG's, Fig. 7) but there are some important differences; for example, the activation free energy of shuttling, $\Delta G^{\ddagger}$, corresponds to the energy required for the macrocycle to move all the way to the new binding site (i.e. includes a contribution for the distance the ring has to move along the track to reach the other station) whereas the $\Delta E^{\ddagger}$, shown here, represents the energy required for the macrocycle to escape the forces exerted through non-covalent binding interactions at a station. The main plot shows the ΔE in terms of the position of the macrocycle along the vector of the thread; the minor plot shows the ΔE in terms of the position of the macrocycle orthogonal to the vector of the thread, illustrating that the thread genuinely behaves as a one-dimensional potential energy surface for the macrocycle

tion. When the macrocycle is on the thread *between* the stations, the hydrogen bonds and other non-covalent binding interactions are broken and no forces are exerted on the macrocycle by the thread (i.e. the gradient of the line is zero).

It is thermal energy which allows the macrocycle to escape these energy wells and explore the full length of the thread. For a population of shuttles at equilibrium, therefore, the macrocycles reside on the different stations and the thread according to a Boltzmann distribution and for a single molecule, these populations correspond to the amount of time the macrocycle spends on each site. In a degenerate shuttle, where the binding energy to each station is the same, while no significant interaction with the thread occurs, the macrocycle spends a negligible amount of time on the thread and splits its time equally between the two stations.

The rate at which the macrocycles escape from each station is given by a standard Arrhenius equation, depending on the depth of the energy well and the temperature. This is not the only factor involved in determining the rate at which macrocycles move between the two stations, however, there must also be some distance-dependent diffusional factor in the function describing the rate of shuttling (Fig. 8).

This phenomenon has been studied in an attempt to link the experimentally determined values of $\Delta G^{\ddagger}$ with a quantum-mechanical description of the shuttling mechanism [66]. Calculation of the wavefunctions for the system shows that an increase in distance between the stations widens the free energy potential well (Fig. 7). As the well widens, it possesses a higher density of states per unit energy (just like the simple "particle in a box" model). The closer to one another the levels are, the more readily thermally populated they are or, in other words, the larger is the partition function and hence $\Delta G^{\ddagger}$.

The nature of the wavefunctions at energies close to the top of the barrier is intriguing, as under this energy regime the maximum probability of finding the macrocycle is in fact over the aliphatic spacer! The shuttling process can therefore be thought of in terms of a function of free energy (Fig. 7) as long as it is remembered that the height of the $\Delta G^{\ddagger}$ barrier is affected by *both* binding strength and distance between the stations. The behavior of the macrocycle is similar to a cart moving along a roller coaster track shaped like the double potential in Fig. 7. At low temperatures, the cart mostly resides on the stations (oscillating with small amplitude in the troughs). As energy approaches the value of the barrier, the cart spends most of its time passing over the barrier. At higher energies still, the cart is most likely to be found at the extremes of its translational motion.

3.4
Translational Molecular Switches: Stimuli-Responsive Molecular Shuttles

With increasing understanding of the nature of the inherent restriction in degrees of freedom in interlocked architectures, came the realization that control of intercomponent positioning was achievable. In shuttles such as **14–16** there are two identical recognition sites (stations) for the macrocycle so that it is equally likely to reside on either—they are degenerate. As we have seen, the rate at which the macrocycle moves between the stations can be regulated by the temperature or, in some cases, solvent composition. A different kind of control, however, can be achieved in stimuli responsive shuttles in which the net location of the macrocycle can be varied using an applied external stimulus. Indeed, as early as 1991 when reporting the first degenerate molecular shuttle [67], Stoddart noted that: *"The opportunity now exists to desymmetrize the molecular shuttle by inserting nonidentical 'stations' along the polyether 'thread' in such a manner that these different 'stations' can be addressed selectively by chemical, electrochemical, or photochemical means and so provide a mechanism to drive the 'bead' to and fro between 'stations' along the 'thread'."*

3.4.1
A Physical Model of Two Binding Site, Stimuli-Responsive Molecular Shuttles

Rotaxanes in which the macrocycle can be translocated between two or more well-separated stations in response to an external signal can, in principle, act as molecular-level mechanical switches. As we have seen, in any rotaxane the macrocycle distributes itself between the available binding sites according to the difference in the macrocycle binding energies and the temperature. If a suitably large difference in macrocycle affinity between two stations exists, the macrocycle resides overwhelmingly in one positional isomer or co-conformation. In stimuli-responsive molecular shuttles, an external trigger is used to chemically modify the system and alter the non-covalent intercomponent interactions such that the second macrocycle binding site becomes energetically more favored, causing translocation of the macrocycle along the thread to the second station (Fig. 9). This may be achieved by addressing either of the stations (destabilizing the initially preferred site or increasing the binding affinity of the originally weaker station). The system can be returned to its original state by using a second chemical modification to restore the initial order of station binding affinities. Performed consecutively these two steps allow the machine to carry out a complete cycle of shuttling motion.

The physical basis for this motion is again best understood by consideration of the potential energy of the macrocycle as a function of its position along the thread (Fig. 10). It is important to appreciate that the external stimulus does not induce directional motion of the macrocycle per se, rather by increasing the binding strength of the less-populated station and/or destabilizing the initially preferred binding site, the system is put out of co-conformational equilibrium. Relaxation towards the new global energy minimum subsequently occurs by *thermally activated* motion of the components, a phenomenon which we recognize as biased Brownian motion. In other words, biased Brownian motion arises from a difference in the activation energies for movement in different directions, *not* from the difference in energy minima. This results in net di-

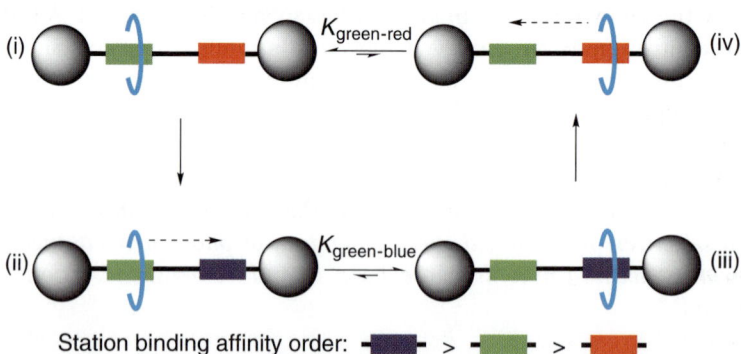

Fig. 9 Translational submolecular motion in a stimuli-responsive molecular shuttle: (i) the macrocycle initially resides on the preferred station (*green*); (ii) a reaction occurs (*red → blue*) changing the relative binding potentials of the two stations such that, (iii), the macrocycle "shuttles" to the now-preferred station (*blue*). If the reverse reaction (*blue → red*) now occurs, (iv), the components return to their original positions

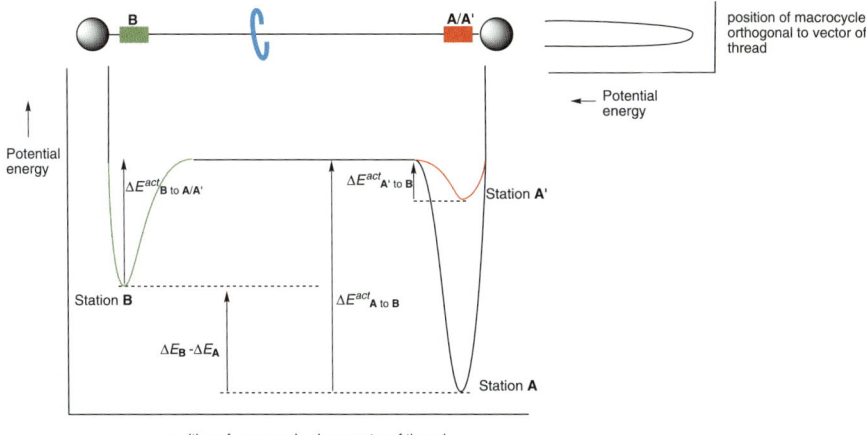

Fig. 10 Idealized potential energy of the macrocycle in a stimuli-responsive molecular shuttle in which one station changes (A′ → A) in response to the stimulus and complicating factors such as folding are ignored. As before (Fig. 8), the potential energy surface shows the effect of the interaction between macrocycle and thread on the energy of the macrocycle. The main plot shows the ΔE in terms of the position of the macrocycle along the vector of the thread; the minor plot shows the ΔE in terms of the position of the macrocycle orthogonal to the vector of the thread

rectional transport (a directional flux) of macrocycles when these barriers are suddenly changed putting the system out of equilibrium.

Given this mode of action, a key requirement is finding ways of generating sufficiently large, long-lived binding energy differences between pairs of positional isomers. A Boltzmann distribution at 298 K requires a $\Delta\Delta E$ (or $\Delta\Delta G$) between translational co-conformers of ~ 2 kcal mol^{-1} for 95% occupancy of one station. Achieving such discrimination in *two* states to form a positionally bistable shuttle (i.e. both $\Delta\Delta E_{A'-B}$ and $\Delta\Delta E_{B-A} \geq 2$ kcal mol^{-1}) by modifying only intrinsically weak, non-covalent binding modes thus presents a significant challenge. In the next sections we outline some of the different stimuli that have been used to bring about a positional change in the macrocycle in amide-based molecular shuttles.

3.4.2
Adding and Removing Protons to Induce Net Positional Change

The first pH-switched amide-based shuttle to exploit [N – H··· anion] hydrogen bonding interactions was recently reported [68]. In [2]rotaxane **18·H** (Scheme 4), formation of the benzylic amide macrocycle is templated by a succinamide station in the thread. However, the thread also contains a cinnamate group. In the neutral form, the cinnamate phenol is a relatively poor hydrogen bonding group and the macrocycle resides on the succinamide station > 95%

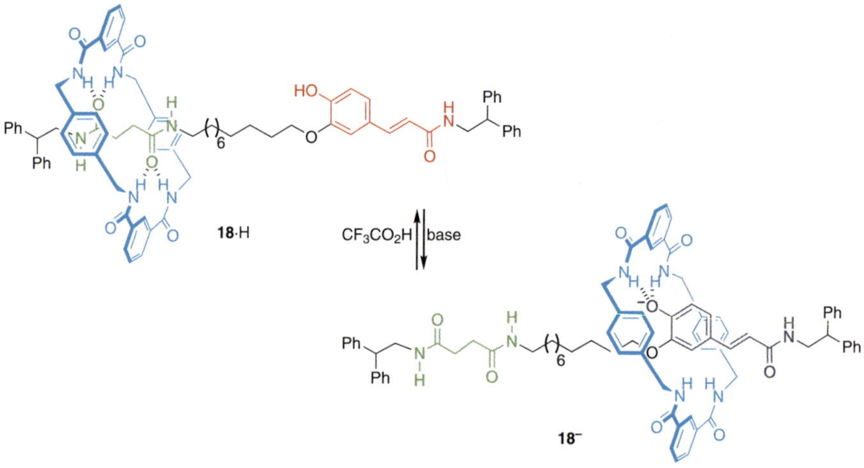

Scheme 4 pH-Switched anion shuttling in a hydrogen bonded [2]rotaxane, **18·H**, in [D$_7$]DMF at rt. Bases that can be used include LiOH, NaOH, KOH, CsOH, Bu$_4$NOH, tBuOK, DBU, Schwesinger's phosphazine P$_1$ base, illustrating the lack of influence on shuttling of the accompanying cation

of the time in most solvents at rt. Deprotonation to give **18$^-$** in [D$_7$]DMF results in the macrocycle binding to the phenolate anion. Reprotonation of the phenol returns the system to its original state. While a wide range of bases (with a variety of counterions) proved efficacious, shuttling was found to be extremely solvent dependent. Hydrogen bond mediated systems usually perform best in "non-competing" solvents—those with low hydrogen bond basicity. Yet, when the deprotonation of **18·H** is carried out in CDCl$_3$ or CD$_2$Cl$_2$, a change of position of the macrocycle does not occur. Rather, an intramolecular folding event occurs to allow the phenolate to hydrogen bond with the macrocycle while it remains on the succinamide station. This solvent dependence presumably arises because the phenolate can only satisfy the hydrogen bonding requirements of one isophthalamide unit in the macrocycle. The presence of a hydrogen bond accepting solvent such as DMF can therefore compensate for this loss of stabilization. The strength of binding to the anion is illustrated by the fact that the shuttling process continues to occur even in CD$_3$CN—only a moderate hydrogen bond acceptor and weaker than the amide groups of the thread. Shuttling is unaffected by the nature of the accompanying cation or the addition of up to 10 equivalents of other anions.

3.4.3
Adding and Removing Electrons to Induce Net Positional Change

[2]Rotaxane **19** contains two potential hydrogen bonding stations for the benzylic amide macrocycle—a succinamide (*succ*) station and a redox-active

3,6-di-*tert*-butyl-1,8-naphthalimide (*ni*) station—separated by a C_{12} aliphatic spacer (Scheme 5) [69, 70].

While the ability of the *succ* station to template formation of the macrocycle is well established, the neutral naphthalimide moiety is a poor hydrogen bond acceptor. To minimize its free energy, the macrocycle in **19** must therefore sit over the succinamide station in non-hydrogen bonding solvents, so co-conformation *succ*-**19** predominates (Scheme 5). In fact, the difference in macrocycle binding affinities is so great that *succ*-**19** is the only translational isomer detectable by ^{1}H NMR in $CDCl_3$, CD_3CN, and $[D_8]THF$, while even in the strongly hydrogen bond-disrupting $[D_6]DMSO$, the macrocycle resides over the *succ* station about half of the time. One-electron reduction of the naphthalimide to the corresponding radical anion, however, results in a substantial increase in electron charge density on the imide carbonyls and a concomitant increase in hydrogen bond accepting ability. In **19**, this reverses the relative hydrogen bonding abilities of the two thread stations so that co-conformation *ni*-**19**$^{-\cdot}$ is preferred in the reduced state. Subsequent

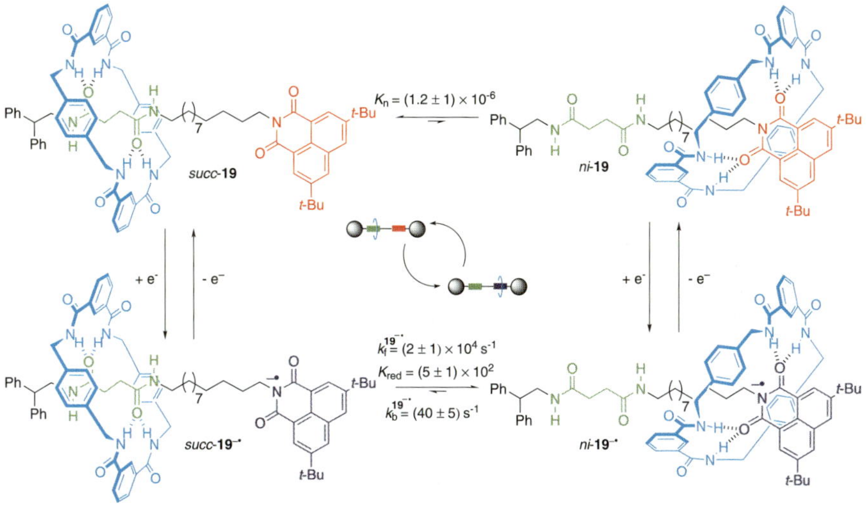

Scheme 5 A photochemically and electrochemically switchable, hydrogen bonded molecular shuttle **19**. In the neutral state, the translational co-conformation *succ*-**19** is predominant as the *ni* station is a poor hydrogen bond acceptor ($K_n = (1.2 \pm 1) \times 10^{-6}$). Upon reduction, the equilibrium between *succ*-**19**$^{-\cdot}$ and *ni*-**19**$^{-\cdot}$ is altered ($K_{red} = (5 \pm 1) \times 10^2$) because *ni*$^{-\cdot}$ is a powerful hydrogen bond acceptor and the macrocycle changes position through biased Brownian motion. Upon re-oxidation, the macrocycle shuttles back to the succinamide station. Repeated reduction and oxidation causes the macrocycle to be displaced forwards and backwards between the two stations. All the values shown refer to cyclic voltammetry experiments in anhydrous THF at 298 K with tetrabutylammonium hexafluorophosphate as the supporting electrolyte. Similar values were determined on photoexcitation and reduction of the ensuing triplet excited state by an external electron donor

re-oxidation to the neutral compound restores the original order of binding site affinities and the shuttle returns to its initial state as co-conformation *succ*-**19**. This process can be stimulated and observed in cyclic voltammetry experiments [70], or alternatively photochemistry can be employed to initiate (through excitation of the naphthalimide group by a nanosecond laser pulse at 355 nm followed by electron transfer from a regenerable external electron donor) and observe (using transient absorption spectroscopy) the change of position of the macrocycle [69]. A number of control experiments proved unequivocally that the dynamic process observed is reversible shuttling of the macrocycle between the stations rather than any other conformational or co-conformational changes [70].

3.4.4
Adding and Removing Covalent Bonds to Induce Net Positional Change

Perhaps surprisingly, the use of covalent bond-forming reactions to bring about positional change in molecular shuttles has been limited to the formation (and breaking) of C – C bonds through Diels–Alder (DA) and retro-Diels–Alder (r-DA) reactions of rotaxane **20** (Scheme 6) [71]. The steric bulk of the DA-adduct displaces the macrocycle to the succinic amide ester station in *Cp*-**20**.

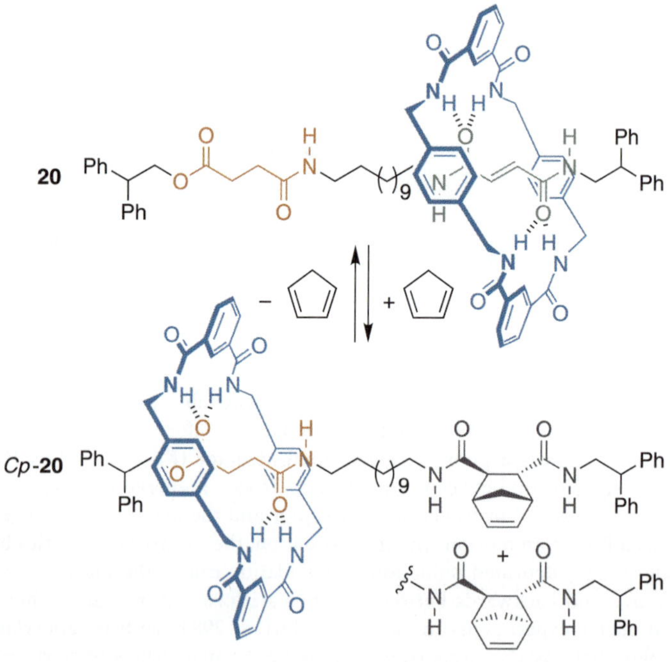

Scheme 6 Shuttling through reversible covalent bond formation. Absolute stereochemistry for *Cp*-**20** is depicted arbitrarily

3.4.5
Changing Configuration to Induce Net Positional Change

As with the control of submolecular motion in covalently bound systems, isomerization processes are an attractive means for controlling shuttling in rotaxanes. Shuttle E/Z-**21** (Scheme 7) employs the interconversion between fumaramide (*trans*) and maleamide (*cis*) isomers of the olefinic unit [72]. Fumaramide moieties are excellent binding sites for benzylic amide macrocycles: the *trans*-olefin fixes the two strongly hydrogen bond accepting amide carbonyls in a close-to-ideal spatial arrangement for interaction with the amide protons from the macrocycle. Two sets of bifurcated hydrogen bonds between macrocycle and thread result. Although a similar hydrogen bonding surface is presented to the macrocycle, binding to the succinamide station results in both a loss in entropy (due to loss of bond rotation) and also one less intracomponent hydrogen bond which cannot be compensated for by the rigid fumaramide station. The result is that only one major positional isomer of E-**21** is observed at room temperature. Photoisomerization of the fumaramide station by irradiation at 254 nm reduces the number of possible intercomponent hydrogen bonds at this station from four to two so that a new co-conformational energy minimum now exists: the macrocycle now sits overwhelmingly on the succinamide station. Unlike the succinamide/naphthalimide system (Scheme 5), this new state is indefinitely stable until a further stimulus is applied—namely thermal or chemical re-isomerization of the maleamide unit back to fumaramide, thus restoring the shuttle to its initial state.

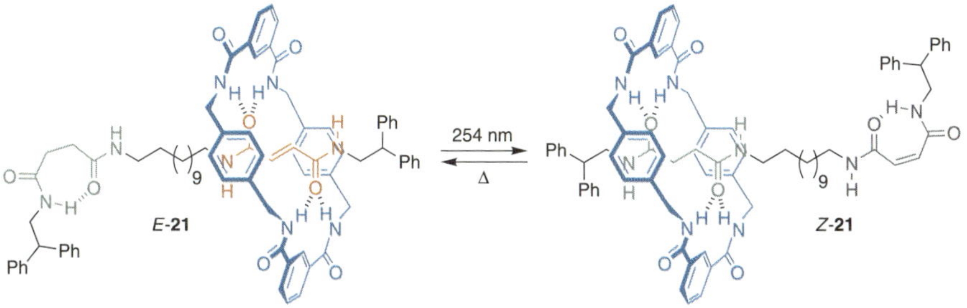

Scheme 7 Bistable molecular shuttle E/Z-**21** in which self-binding of the "low affinity" station in each state is a major factor in producing excellent positional discrimination

3.4.6
Entropy-Driven Net Positional Change

Most of the shuttles which exhibit excellent positional discrimination are switched using stimuli such as pH, light, polarity of the environment, or electrochemistry to modify the enthalpy of macrocycle binding to one or both stations. Generally, the effect of temperature is simply to alter the degree of discrimination the macrocycle expresses for the various stations, not to alter the station preference. In [2]rotaxane **22**, however, the macrocycle is actually switched between stations by changing the temperature [73]. In fact, **22** is a *tristable* molecular shuttle: a rotaxane in which the ring can be switched between three different positions on the thread (Scheme 8).

Structurally, **22** is closely related to **21**, the key difference being substitution of the isophthaloyl unit in the macrocycle for a pyridine-2,6-dicarbonyl moiety. In the *E*-**22** form, the macrocycle resides over the strong fumaramide station at all temperatures investigated, as expected. Photoisomerization of *E*-**22** gave the maleamide *Z*-**22** isomer. The ^{1}H NMR spectra of this product showed clearly that shuttling away from the maleamide station had occurred, but the nature of the product was highly temperature dependent. At elevated temperatures (308 K) the expected *succ-Z*-**22** co-conformation was observed; the macrocycle spends nearly all its time over the succinamide station. At lower temperatures, however, the macrocycle occupies neither the succinamide nor the maleamide units, but it is the alkyl chain which exhibits the spectroscopic shifts indicative of encapsulation by the ring. This suggests that the thread adopts an S-shaped conformation with the macrocycle binding to one amide of each station (*dodec-Z*-**22**, Scheme 8). Unlike [2]rotaxane **21**, photochemical *cis*→*trans* isomerization was effective (irradiation of *Z*-**22** at 312 nm gave a photostationary state of > 95 : 5 *E* : *Z* compared to ∼ 45 : 55

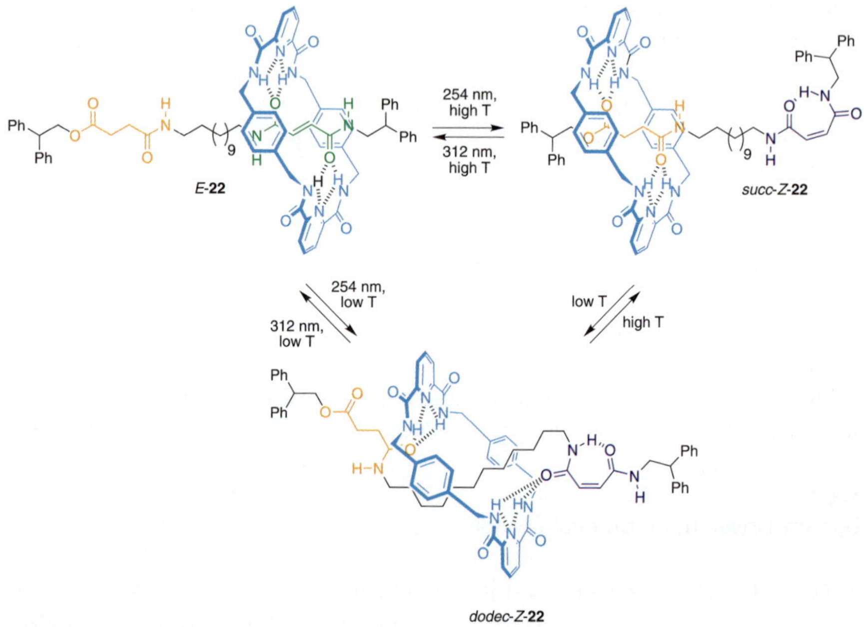

Scheme 8 A tristable molecular shuttle **22**

under the same conditions for **21**) so that no heating is necessary for this step and all the interconversions shown in Scheme 8 are possible.

The origin of this temperature-switchable effect is presumably the large difference in entropy of binding ($\Delta S_{binding}$) to the succinamide and alkyl chain stations which allows the $T\Delta S_{binding}$ term to have a significant overall impact on $\Delta G_{binding}$ as temperature is varied. In the *succ-Z-***22** co-conformation, the macrocycle forms two strong hydrogen bonds with an amide carbonyl and two, significantly weaker, bonds to the ester carbonyl. The *dodec-Z-***22** co-conformation, however, allows formation of four strong hydrogen bonds to amide carbonyls making it enthalpically favored by ~ 2 kcal mol^{-1}. At low temperatures, where the effects of the entropy term are less significant, the molecule therefore adopts the *dodec-Z-***22** co-conformation. At higher temperatures, the increased contribution from the $T\Delta S_{binding}$ term requires that the molecule adopt the more entropically favorable *succ-Z-***22** co-conformation in order to minimize its energy.

This effect seems to be quite structure specific—no other shuttles in this series have shown temperature dependent co-conformational preference. If suitable systems can be successfully designed however, entropy-driven temperature control of positional isomerism could prove a useful addition to the expanding strategies for controlling submolecular motions [74].

Many of the shuttling examples described show remarkable degrees of control over submolecular fragment positioning and dynamics. They utilize a number of different stimuli-induced processes to effect macrocycle shuttling over large amplitudes (up to ~ 15 Å in the case of **14–16** and **18–22**) and operate over a range of timescales (a complete shuttling cycle in **19** is over in ~ 100 μs while, in **21**, both states are indefinitely stable and the time for a full cycle is infinitely variable). Yet it must be remembered that all such shuttles exist as an equilibrium of co-conformations and it is simply the position of the equilibrium that is varied.

3.5
Controlling Rotational Motion: Ring Pirouetting in Rotaxanes

Control of macrocycle pirouetting in rotaxanes presents two challenges—frequency of random pirouetting and directionality. The former has been achieved through temperature, structure, electric fields, light, and solvent effects; the latter has yet to be demonstrated.

Alternating current (a.c.) electric fields are an ideal stimulus with which to control submolecular dynamics for many applications. For two benzylic amide-based hydrogen bonded [2]rotaxanes (**23** and **24**, Fig. 11) it was observed that application of a.c. fields of around 50 Hz resulted in unusual Kerr effect responses which are unique to the interlocked architecture (i.e. are not observed for either of the components alone) [75]. Decreasing the field strength had the same effect as increasing temperature: enhancement of the response; indicating

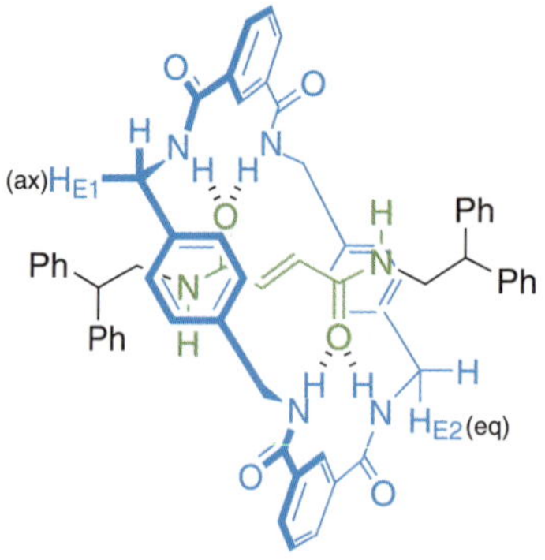

Fig. 11 [2]Rotaxanes **23** and **24** in which the rate of pirouetting is controlled by an alternating current electric field

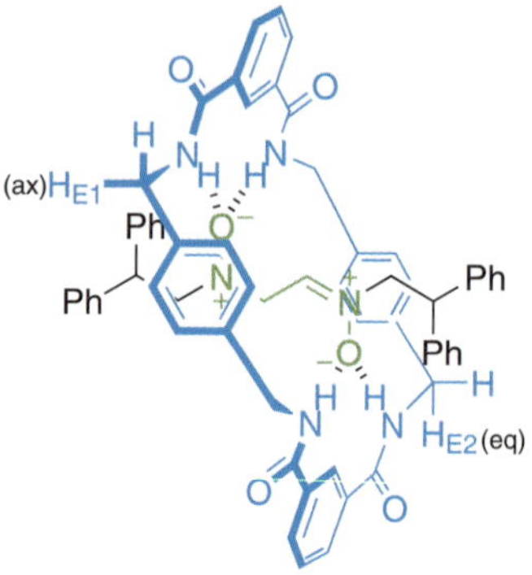

that the underlying phenomenon is addressable by these two stimuli. Both VT NMR experiments and molecular modeling suggested that macrocycle pirouetting is the only possible dynamic process on this timescale for these structures. The experimental and theoretical studies also predict the slightly more complex Kerr effect response observed experimentally for **23** compared to **24**. It is believed that application of an a.c. electric field therefore attenuates macrocycle pirouetting in **18** and **19**—the extent of the dampening can be varied with the strength of the applied field; even modest fields of ~ 1 V cm^{-1} produce rate reductions of 2–3 orders of magnitude.

An alternative strategy for affecting pirouetting rates would be to apply some stimulus which directly alters the structure or electronics of the thread or macrocycle so as to adjust the strength of intercomponent interactions. This has been achieved for the fumaramide-based [2]rotaxanes E/Z-**25**–**27** (Scheme 9), which are closely related to **23** [76]. The decrease in intercomponent binding affinity on photoisomerization of the fumaramide units in E-**25**–**27** to the *cis*-maleamide isomers gives a huge increase in the rate of pirouetting of more than 6 orders of magnitude. The switching process is also reversible; subjecting the maleamide rotaxanes to heat or various chemical stimuli results in re-isomerization to the more thermally stable *trans*-olefin isomers, with accompanying reinstatement of the strong hydrogen bonding network.

In all of the above systems, the rate of Brownian pirouetting is influenced by applying external stimuli, the direction of these random motions are not affected. It has incorrectly been suggested [77] that unidirectional pirouetting could result from simply derivatizing rotaxanes with chiral or knotted stoppers. Directional rotation can only result from an external energy source being used (more than once in order to achieve circumrotation) to drive a system temporarily away from equilibrium. Under those circumstances, simple 2D asymmetry would be sufficient to induce directional motion.

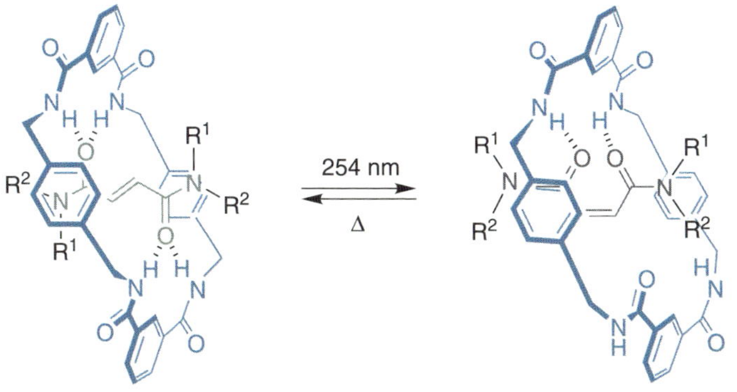

E-25 R¹ = R² = CH₂CO₂CH₂Ph Z-25
E-26 R¹ Me, R₂ = CH₂CHPh₂   Z-26
E-27 R¹ = H, R² = CH₂CHPh₂ Z-27

Scheme 9 Photoisomerization of [2]rotaxanes E-25-27 which results in pirouetting rate enhancements of up to 6 orders of magnitude. The reverse process Z → E isomerization can be effected by heating a 0.02 M solution of the Z-rotaxanes at 400 K or generating bromine radicals (cat. Br₂, hv 400 nm) or via reversible Michael addition of piperidine (rt, 1 h)

3.6
Controlling Rotational Motion in Catenanes

3.6.1
Two-Way and Three-Way Catenane Positional Switches

The fundamental principles of controlling shuttling in rotaxanes and rotation in catenanes are the same. For example, homocircuit [2]catenane **28**, acts somewhat like the two station degenerate shuttle **14** [78]. In halogenated solvents such as CDCl$_3$, the two macrocycles interact through hydrogen bonding between their aromatic-1,3-diamide groups, resulting in a "host-guest" relationship in which each (constitutionally identical) ring adopts a different conformation and exists in a different chemical environment (*amide-endo-***28**, Scheme 10). Pirouetting of the two rings interconverts this host-guest relationship and is fast on the NMR timescale in CDCl$_3$ at rt. In a hydrogen bond-disrupting solvent such as [D$_6$]DMSO, however, the preferred co-conformation has the amides exposed on the surface where they can interact with the surrounding medium, while the hydrophobic alkyl chains are buried in the middle of the molecule (*amide-exo-***28**, Scheme 10). Of course, with the disruption of the main non-covalent interactions between the two rings, the frequency of movements is also greatly increased in polar media.

As with rotaxanes, stimuli-induced structural changes can be used to alter the rates of the random intercomponent motions in catenanes. For example, electrochemical reduction of benzylic amide [2]catenane **3** completely halts

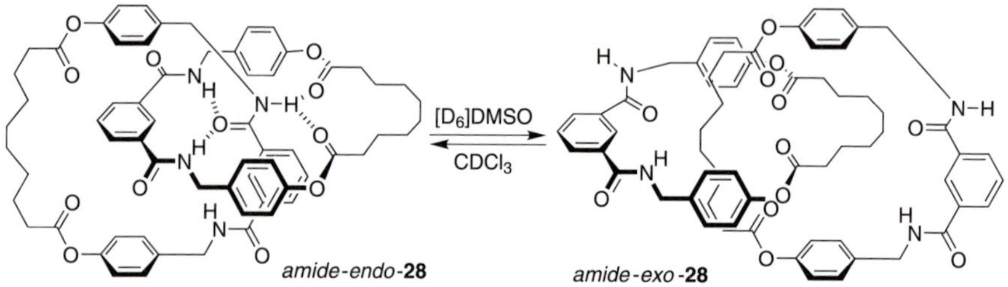

Scheme 10 Translational isomerism in an amphiphilic benzylic amide [2]catenane **28**

the circumrotational process due to formation of an intramolecular covalent bond between the two rings [79].

Sequential movement of one macrocycle between *three* stations on a second ring requires independent switching of the affinities for two of the units so as to change the relative order of binding affinities, as shown schematically in Scheme 11 [80].

In [2]catenane **29** (Fig. 12), this is achieved by employing two fumaramide stations with differing macrocycle binding affinities, one of which (station **A**, green) is located next to a benzophenone unit. This allows selective, photosensitized isomerization of station **A** by irradiation at 350 nm, before direct photoisomerization of the other fumaramide station (station **B**, red) at 254 nm. Station **B**, being a methylated fumaramide residue, has a lower affinity for the macrocycle than station **A**. The third station (station **C**, orange)— a succinic amide ester—is not photoactive and is intermediate in macrocycle

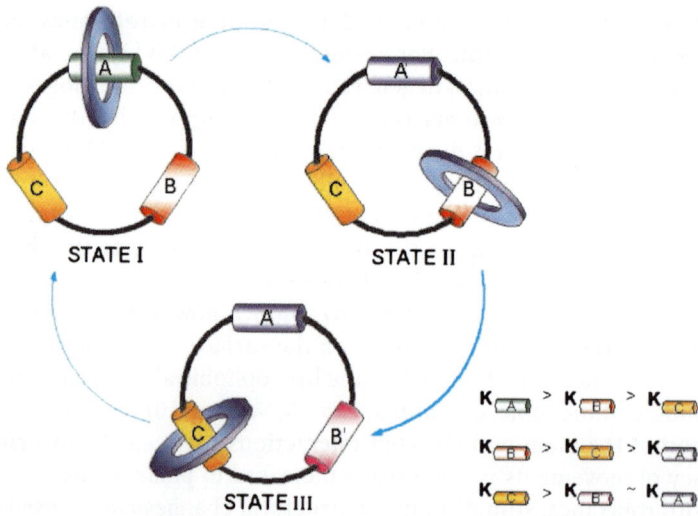

Scheme 11 Stimuli-induced sequential movement of a macrocycle between three different binding sites in a [2]catenane

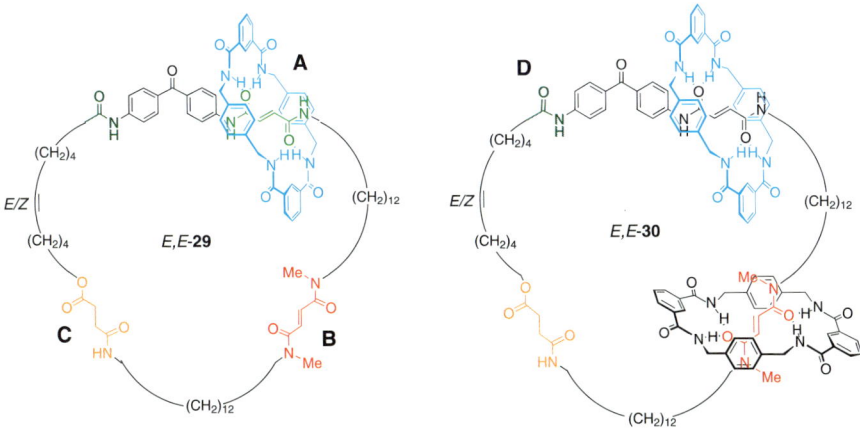

Fig. 12 [2]Catenane **29** and [3]catenane **30** shown as their *E,E*-isomers

binding affinity between the two fumaramide stations and their maleamide counterparts. A fourth station, an isolated amide group (shown as **D** in *E,E*-**30**) which can make fewer intercomponent hydrogen bonding contacts than **A**, **B**, or **C**, is also present but only plays a significant role in the behavior of the [3]catenane.

Consequently, in the initial state (state I, Scheme 11), the small macrocycle resides on the green, non-methylated fumaramide station of [2]catenane **29**. Isomerization of this station (irradiation at 350 nm, green → blue) destabilizes the system and the macrocycle finds its new energy minimum on the red station (state II). Subsequent photoisomerization of this station (irradiation at 254 nm, red → pink) means the macrocycle must now move onto the succinic amide ester unit (orange, state III). Finally, heating the catenane (or treating it with photo-generated bromine radicals or piperidine) results in isomerization of both the Z-olefins back to their *E*-forms (pink → red and blue → green) so that the original order of binding affinities is restored and the macrocycle returns to its original position on the green station.

The ^{1}H NMR spectra for each diastereomer show excellent positional integrity of the small macrocycle in this three-way switch at all stages of the process, but the rotation is not directional—over the complete sequence of reactions, an equal number of macrocycles go from **A**, through **B** and **C**, back to **A** again in each direction.

3.6.2
Directional Circumrotation: A [3] Catenane Rotary Motor

Remarkable directional rotary motion has been produced in covalently linked structures in recent years [81–91]. In order to bias the direction the macrocycle takes from station to station in a catenane such as **29**, temporary barriers are required at each stage to restrict Brownian motion in one particular direction and bias the path taken by the macrocycle from station to station. Such a situation is intrinsically present in [3]catenane **30** (Scheme 12) [80]. Irradiation at 350 nm of *E,E*-**30** causes counter-clockwise (as drawn) rotation of the light blue macrocycle to the succinic amide ester (orange) station to give *Z,E*-**30**. Isomerization (254 nm) of the remaining fumaramide group causes the other (purple) macrocycle to relocate to the single amide (dark green) station (*Z,Z*-**25**) and, again, this occurs counter-clockwise because the clockwise route is blocked by the other (light blue) macrocycle. This "follow-the-leader" process, each macrocycle in turn moving and then blocking a direction of passage for the other macrocycle, is repeated throughout the sequence of transformations shown in Scheme 12. After three diastereomer interconversions, *E,E*-**30** is again formed but 360° rotation of each of the small rings has not yet occurred, they have only swapped places. Complete unidirectional ro-

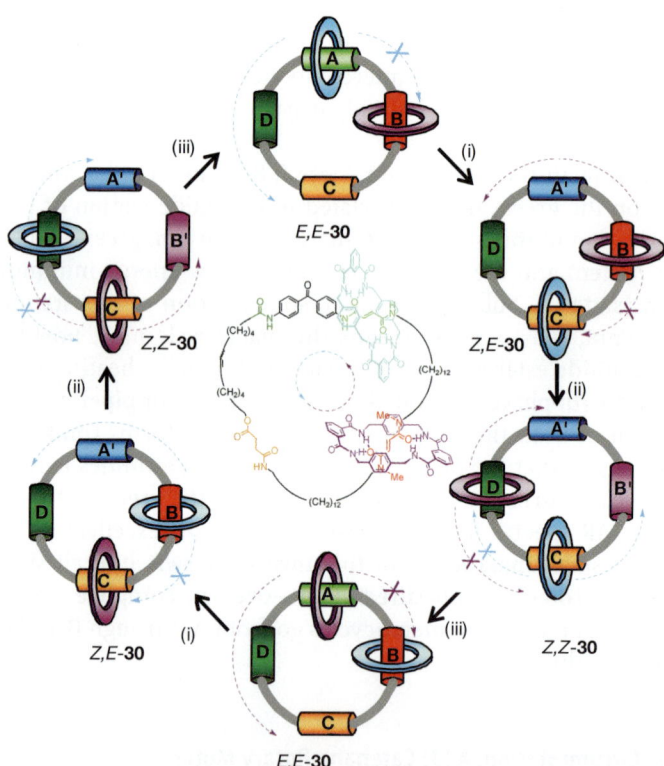

Scheme 12 Stimuli-induced unidirectional rotation in a four station [3]catenane **30**. (i) 350 nm, CH$_2$Cl$_2$, 5 min, 67%; (ii) 254 nm, CH$_2$Cl$_2$, 20 min, 50%; (iii) 100 °C, C$_2$H$_2$Cl$_4$, 24 h, ~ 100%; or catalytic ethylenediamine, 50 °C, 48 h, 65%; or catalytic Br$_2$, 400–670 nm, CH$_2$Cl$_2$, – 78 °C, 10 min, ~ 100%

tation of both small rings occurs only after the synthetic sequence (i)–(iii) has been completed twice.

3.6.3
Selective Rotation in Either Direction: A [2] Catenane Reversible Rotary Motor

Catenane **30** rotates directionally solely through the biasing of random Brownian motion. Over the past decade, a number of theoretical formalisms have been developed using non-equilibrium statistical physics which explain how various types of random fluctuation-driven transport can occur [7, 9–18]. Underlying each of these Brownian ratchet or motor mechanisms are three components: (i) a randomizing element; (ii) an energy input to avoid falling foul of the Second Law of Thermodynamics; and (iii) asymmetry in the energy or information potential in the dimension in which the motion occurs. Such ratchet mechanisms not only account for the general principles behind biological motors but have also been successfully applied to the development of transport and separation devices for mesoscopic particles and macromolecules, microfluidic pumping, the photo-alignment of liquid crystals, and quantum and electronic applications. Accordingly it appeared to us that a consideration of such physical mechanisms could aid the understanding of how to direct intramolecular rotations within chemical structures and we applied these ideas to a catenane architecture [92].

A flashing ratchet is a particular type of energy ratchet mechanism [11], a classic example of which consists in physical terms (Fig. 13) of an asymmetric potential energy surface (a periodic series of two different minima and two different maxima) along which a Brownian particle is directionally transported by sequentially raising and lowering each set of minima and maxima by changing the potential (for example, with an oscillating electric field and a charged particle). *The key to visualizing how the principles of such an energy ratchet can be applied to a catenane architecture is not to consider the whole catenane as a molecular machine, but rather to think of one macrocycle as a motor that transports a substrate—the other ring—directionally around itself!* In its simplest form this results in a [2]catenane such as **31** (Fig. 14) which is able to directionally rotate the smaller ring about the larger one in response to a series of chemical reactions.

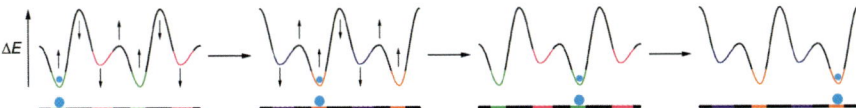

Fig. 13 A flashing energy ratchet mechanism for Brownian particle transport along an oscillating potential energy surface

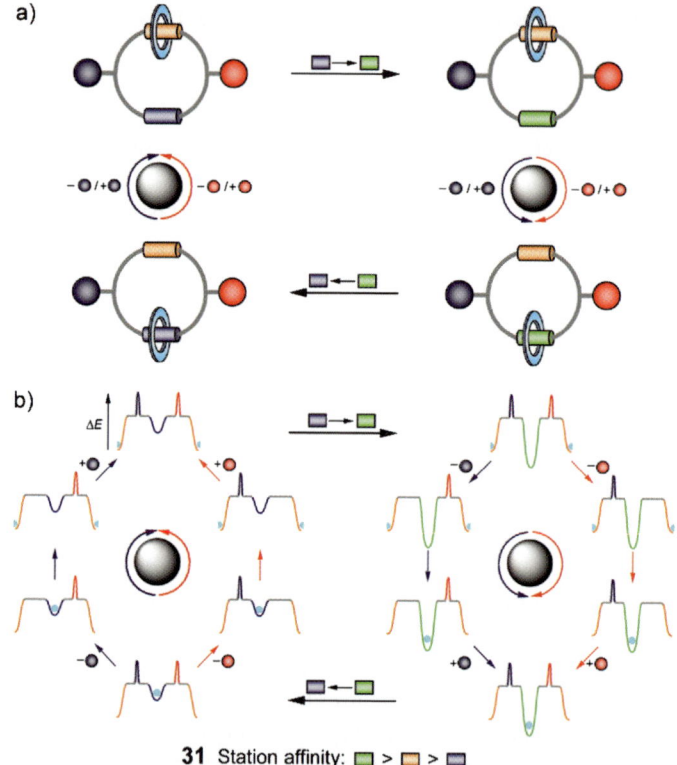

Fig. 14 a Schematic illustration and **b** potential energy surface for the blue ring in a minimalist [2]catenane rotary molecular motor, **31**

This concept was realized in chemical terms through the synthesis and operation of catenane **32** (Scheme 13) [92]. Net changes in the position or potential energy of the smaller ring were sequentially achieved by: (i) photoisomerization to the maleamide (→*mal-Z*-**32**); (ii) de-silylation/re-silylation (→*succ-Z*-**32**); (iii) re-isomerization to the fumaramide (→*succ-E*-**32**); and finally, (iv) de-tritylation/re-tritylation to regenerate *fum-E*-**32**, the whole reaction sequence producing a net clockwise (as drawn in Scheme 13) circumrotation of the small ring about the larger one. Exchanging the order of steps (ii) and (iv)—i.e. employing steps (v) and (vi) instead—produced an equivalent counter-clockwise rotation of the small ring.

Biological motors are obviously too complex for the thermodynamic function of individual amino acid movements to be unraveled in detail. In contrast, the simplicity of **32** and the minimalist nature of its design allows insight into the fundamental role each part of the structure plays in the operation of the rotary machine. The various chemical transformations perform two different functions: one pair (the linking/unlinking reactions—steps (ii) and (iv) or (v) and (vi)) modulates whether the small macrocycles can be

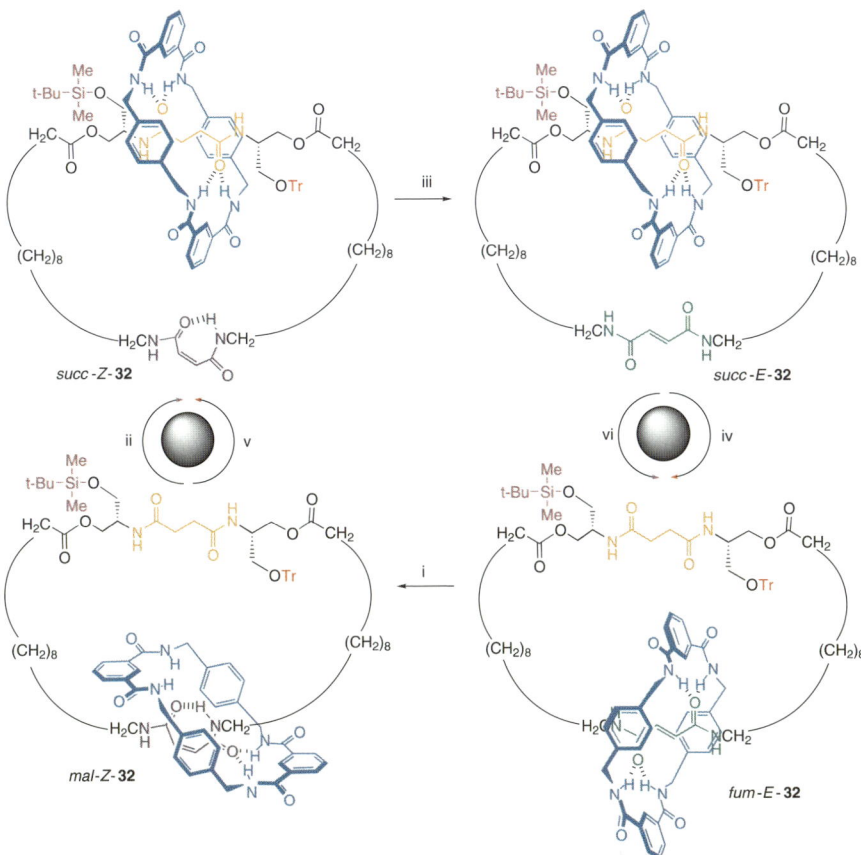

Scheme 13 A reversible [2]catenane rotary motor **32**. i) hν 254 nm, 5 min., 50%; ii) TBAF, 20 min then cool to −78 °C and add 2,4,6-collidine, TBDMSOTf, 1 h, overall 61%; iii) piperidine, 1 h, ∼100%; iv) Me$_2$S·BCl$_3$, −10 °C, 15 min then cool to −78 °C and add 2,4,6-collidine, TrOTf, 5 h, overall 63%; v) Me$_2$S·BCl$_3$, −10 °C 10 min and then TrCl, Bu$_4$NClO$_4$, 2,4,6-collidine, 16 h, overall 74%; vi) TBAF, 20 min then cool to −10 °C and add 2,4,6-collidine, TBDMSOTf, 40 min, overall 76%

exchanged between the two binding sites on the big ring or not (i.e. allow the small macrocycle to reach positional equilibrium and become statistically balanced between the two binding sites according to a Boltzmann distribution); the second pair (balance-breaking reactions—steps (i) and (iii)) isomerize the olefin station (either $E \rightarrow Z$ or $Z \rightarrow E$), switching its binding affinity for the small macrocycle either on or off. By changing the relative binding affinities of the two stations in the large ring, each balance-breaking stimulus provides a driving force for re-distribution of the small ring if it is able to move between the binding sites. In other words, the balance-breaking reactions control the *thermodynamics* and *impetus for net transport* by biased

Brownian motion; the linking/unlinking reactions largely control the relative *kinetics* and *ability to exchange*. Raising a kinetic barrier also "ratchets" transportation, allowing the statistical balance of the small ring to be subsequently broken without reversing the preceding net transportation sequence. Lowering a kinetic barrier allows "escapement" of a ratcheted quantity of rings in a particular direction.

To obtain 360° rotation of the small ring about the large ring, the four sets of reactions must be applied in one of two sequences, each taking the form: first a balance-breaking reaction; then a linking/unlinking step; then the second balance-breaking reaction; finally, the second linking/unlinking step. The direction of net rotation is determined solely by the way the balance-breaking and linking/unlinking steps are paired—an external input of information. The sense of rotation is not affected by any of the intrinsic requirements—2D symmetry breaking, energy input, thermal bath—for directional transport. The efficiency or yields of the reactions—or the position of the ring at any stage (even if the machine makes a "mistake")—are immaterial to the direction in which net motion occurs, as long as the reactions continue to be applied in the same sequence. Although reversing the sequence of the four steps changes the pairings and so rotates the small ring in the opposite direction, reversing the entire sequence of six chemical reactions does not, because linking-unlinking operations are not commutative.

Catenane **32** demonstrates that mechanisms formulated from non-equilibrium statistical mechanics can be successfully used to design synthetic molecular motors. In turn, the analysis of this deceptively simple molecule—particularly the separation of the kinetic and thermodynamic requirements for detailed balance—provides experimental insight into how an energy input is essential for directional rotation of a submolecular fragment by Brownian motion. Even though no net energy is used to power the motion, there has to be some processing of chemical energy for net rotation to be directional over a statistically significant number of molecules; a requirement that is absent if the equivalent motion is non-directional. The amount of energy conversion required to induce directionality has an intrinsic lower limit, corresponding to the binding energy difference of the fumaramide and maleamide binding sites, the same value that determines the directional efficiency of rotation and the maximum amount of work the motor can theoretically perform in a single cycle. The link between information and thermodynamics has haunted physics for nearly 150 years [93]. The factors that determine the sense of rotation in **32** (relying on the sequence, not energetics, of balance-breaking and linking/unlinking steps), together with the requirement for finite energy conversion to heat for directional rotation in circumstances when no work is done against an external force, illustrates how fundamental interplay between informational and thermodynamic laws governs directional Brownian rotation in molecular structures.

4
Property Effects Using Amide-Based Synthetic Molecular Machines

Whatever the future application of synthetic molecular machines may be, it is clear practical devices will require that the machine and its components are able to interact with the macroscopic world, either directly or through further interactions with other molecular-scale devices. Stimuli responsive molecular shuttles offer a generic approach which could be taken to create mechanical molecular switches for a variety of distance dependent properties. Suitable functionalization of the macrocycle and one end of the thread can lead to molecular switches that can change a variety of properties in response to a particular switching stimulus (Fig. 15) [94].

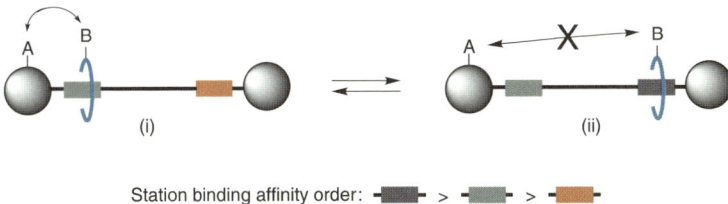

Fig. 15 Exploiting a well-defined, large amplitude positional change to trigger property changes. (i) A and B interact to produce a physical response (fluorescence quenching, specific dipole or magnetic moment, NLO properties, color, creation/concealment of a binding site or reactive/catalytic group, hydrophobic/hydrophilic region, etc.); (ii) moving A and B far apart mechanically switches off the interaction and the corresponding property effect

4.1
Switching On and Off Induced Circular Dichroism with a Molecular Shuttle

In a study of chiral dipeptide [2]rotaxanes, it had been shown that the presence of an intrinsically achiral benzylic amide macrocycle near to the chiral center can induce an asymmetric response in the aromatic ring absorption bands [95]. The induced circular dichroism (ICD) effect is strongest in apolar solvents when intercomponent interactions are maximized and the chirality is transmitted from the amino acid asymmetric center on the thread, via the achiral macrocycle to the aromatic rings of the achiral C-terminal stopper on the thread. These observations led to the design of the chiroptical molecular shuttle *E/Z*-**33** (Scheme 14) [96]. Unlike chiroptical switches in which the presence or handedness of chirality is intrinsically altered, *E/Z*-**33** remains chiral with the same handedness throughout; it is the *expression* of that chirality that is altered. In the *E*-**33** form, the macrocycle is held over the fumaramide binding site, far from the chiral center of the peptidic station. Correspondingly, the circular dichroism response is zero. In the *Z*-**33**

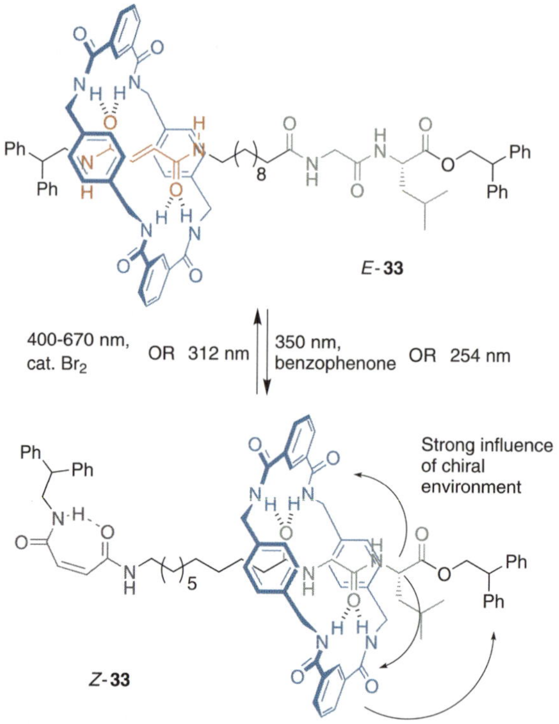

Scheme 14 Chiroptical switching in [2]rotaxane-based molecular shuttle E/Z-**33**

isomer, however, the macrocycle resides on the peptide station close to the L-Leu residue and a strong (– 13 k deg cm^2 dmol^{-1}), negative ICD response is observed [96]. Preparatively, the $E \rightarrow Z$ isomerization is most efficiently carried out by irradiation at 350 nm in the presence of a benzophenone sensitizer (photostationary state 70 : 30 $Z : E$), while the $Z \rightarrow E$ transformation can be achieved almost quantitatively by irradiation at 400–670 nm in the presence of catalytic Br$_2$.

4.2
Switching On and Off Fluorescence with a Molecular Shuttle

A similar approach has been used to make a molecular shuttle switch for fluorescence, E/Z-**34** (Scheme 15) [94]. This system also relies on the photoswitchable fumaramide/maleamide station but attached to the intermediate-affinity dipeptide station is an anthracene fluorophore, while the macrocycle now contains pyridinium units—known to quench anthracene fluorescence by electron transfer. In both the free thread and E-**34**, strong fluorescence (λ_{exc} = 365 nm) is observed, while shuttling of the macrocycle onto the glycylglycine station in Z-**34** quenches this emission almost completely. At the maximum of E-**34** emission

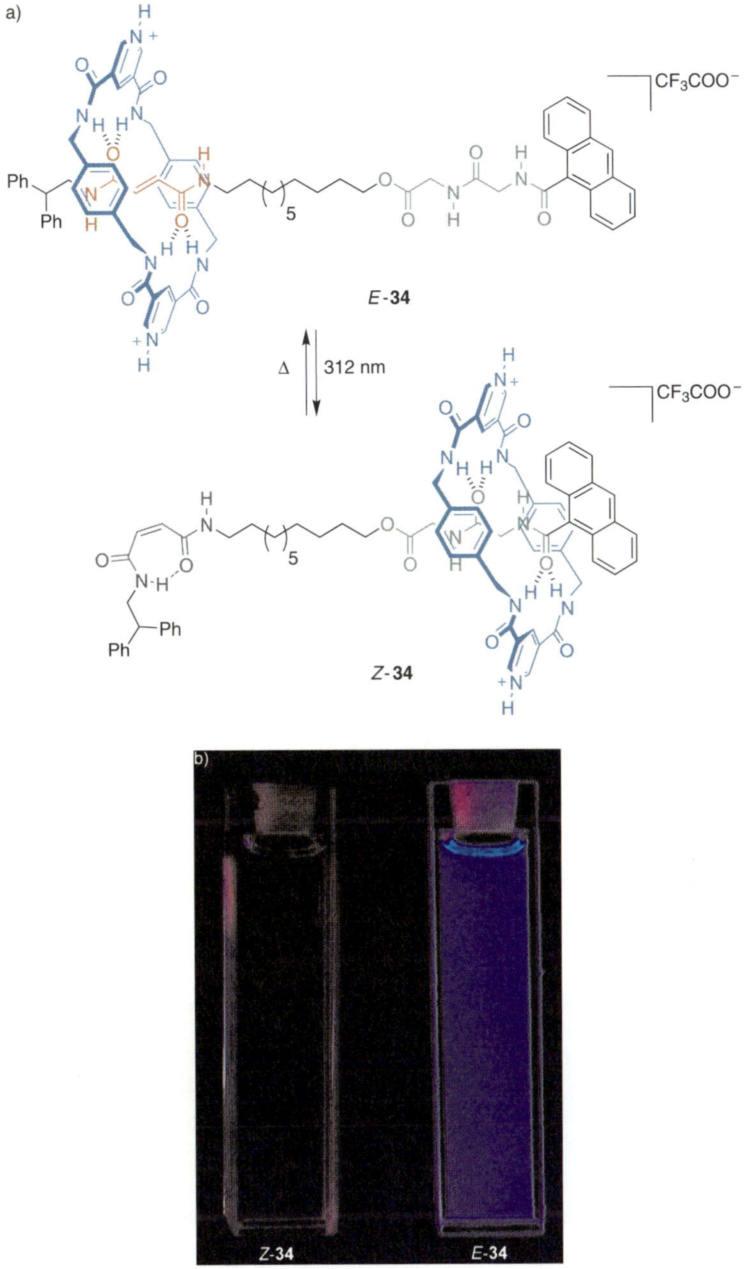

Scheme 15 A fluorescent molecular switch based on [2]rotaxane molecular shuttle E/Z-34. **a** Interconversion between fluorescent E-34 and non-fluorescent Z-34; **b** images of cuvettes containing solutions of Z-34 and E-34 respectively (0.8 μM, CH_2Cl_2) demonstrating the clearly visible difference in fluorescence intensity. The photographs were taken while illuminating with UV light (365 nm)

(λ_{max} = 417 nm) there is a remarkable 200 : 1 difference in intensity between the two states—strikingly visible to the naked eye (Scheme 15b).

The same principles have been used to create switches that function in polymer films [97]. Patterns visible to the naked eye were generated using an environment-switchable shuttle covalently derivatized with poly(methyl methacrylate) (PMMA), **35** (Scheme 16 and Fig. 16). A polymer film INHIBIT logic gate based on a combination of control of submolecular positioning and chemical modification (protonation) was also demonstrated (**36/36·2H$^+$**, Scheme 16 and Fig. 17) [97].

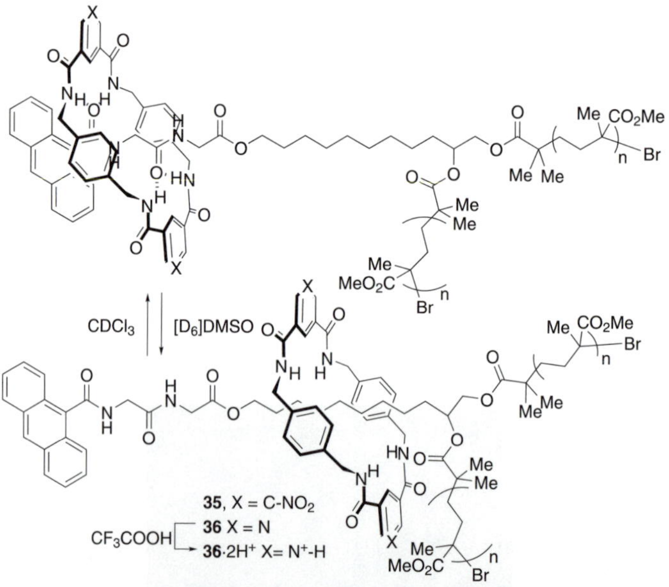

Scheme 16 Polymeric environment-switchable molecular shuttles **35** and **36/36·2H$^+$**

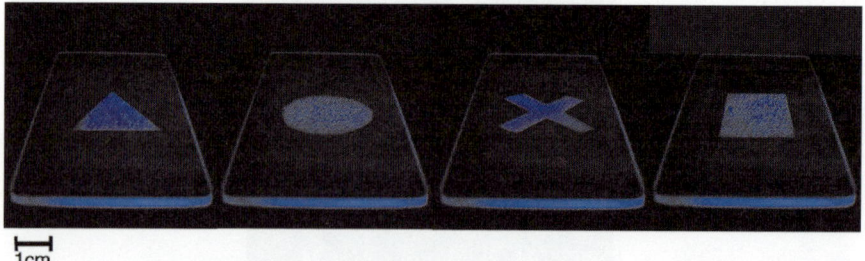

Fig. 16 Images obtained by casting films of polymer **35** on quartz slides, covering them with an aluminum mask and exposing the unmasked area to dimethylsulfoxide vapors for 5 minutes. The photographs were taken while illuminating the slides with UV light (254–350 nm). The symbols of Sony Playstation™ are illustrative of the types of patterns that can be created

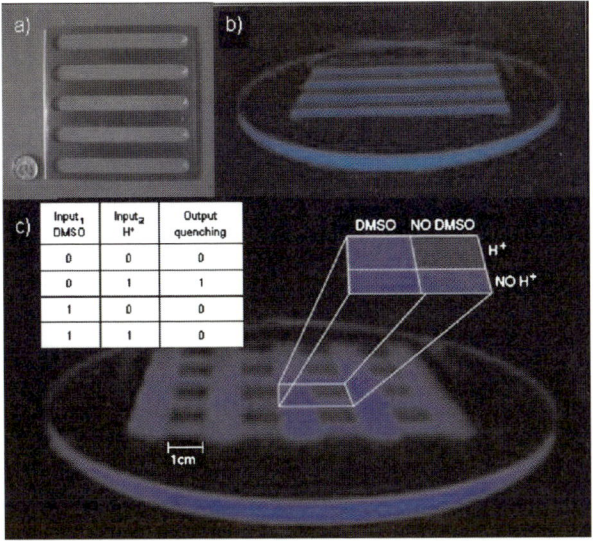

Fig. 17 A molecular shuttle Boolean logic gate that functions in a polymer film. **a** Aluminium grid used in the experiment. The coin shown for scale is a UK 5p piece. **b** Pattern generated when films of **36** were exposed to trifluoroacetic acid vapor for 5 minutes through the aluminium grid mask. **c** Criss-cross pattern obtained by rotating the aluminium grid 90° and exposing the film shown in (**b**) to DMSO vapor for a further 5 minutes. Only regions exposed to trifluoroacetic acid but not to DMSO are quenched. The truth table for an INHIBIT logic gate is shown in the *inset*. The photographs of the slides were taken in the dark while illuminating with UV light (254–350 nm)

4.3
Rotaxane-Based Photoresponsive Surfaces and Macroscopic Transport by Molecular Machines

Perhaps the most dramatic illustration of the potential of controlled molecular level motion in this type of system is a recent demonstration of the creation and utility of photoresponsive surfaces based on rotaxanes (Schemes 17 and 18, Fig. 18) [98]. The millimeter scale directional transport of diiodomethane across a surface (Fig. 18) was achieved using the biased Brownian motion of the components of a stimuli-responsive rotaxane **37** (Scheme 17) to expose or conceal fluoroalkane residues and thereby modify surface tension. The collective operation of a monolayer of the molecular shuttles attached to a self-assembled monolayer of 11-mercaptoundecanoic acid (11-MUA) on Au(111) (Scheme 18) was sufficient to power the movement of a microliter droplet of diiodomethane up a twelve degree incline (Fig. 18e–h).

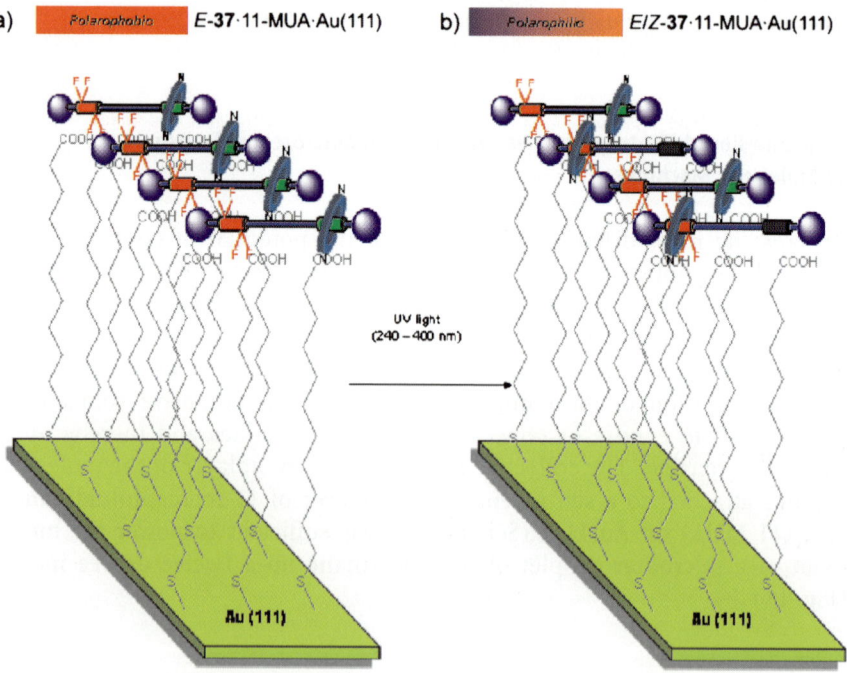

Scheme 17 Stimuli-induced positional change of the macrocycle in a fluorinated molecular shuttle, $37 \cdot 2H^+$. (i) 254 nm, CH_2Cl_2, 5 min, 50%; (ii) piperidine, CH_2Cl_2, rt, 2 h then CF_3CO_2H, 100% or 115°, $C_2H_2Cl_4$, 24 h, 90%

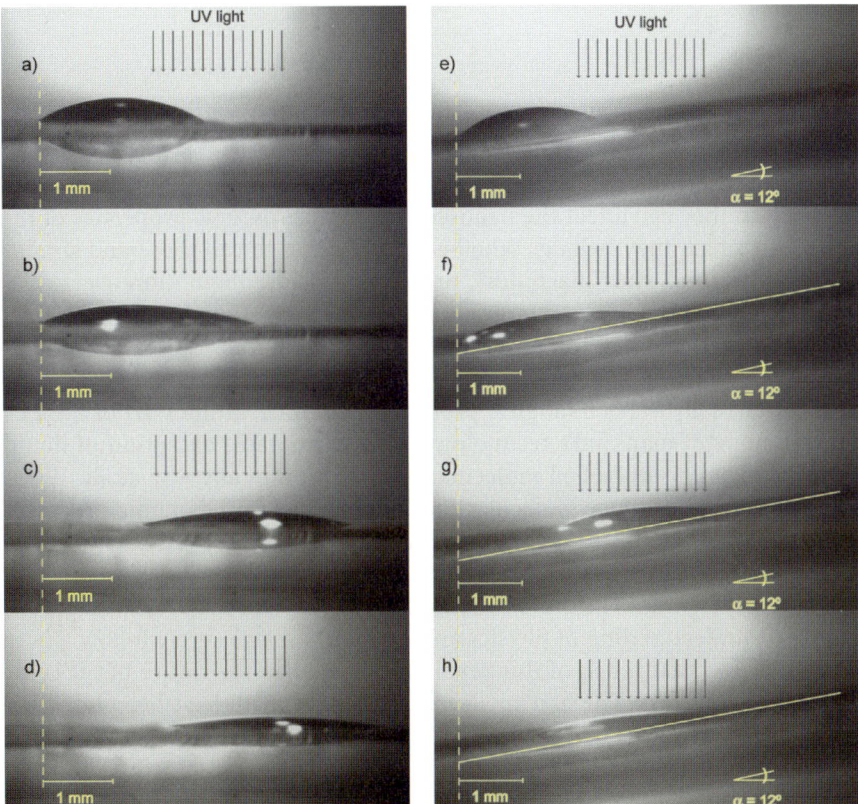

Fig. 18 Lateral photographs of light-driven directional transport of a 1.25 μl diiodomethane drop across the surface of a E-37·11-MUA·Au(111) substrate on mica arranged flat (**a–d**) and up a twelve degree incline (**e–h**). **a** Before irradiation (pristine E-37). **b** After 215 s of irradiation (20 s prior to transport) with UV light in the position shown (the right edge of the droplet and the adjacent surface). **c** After 370 s of irradiation (just after transport). **d** After 580 s of irradiation (at the photostationary state). **e** Before irradiation (pristine E-37). **f** After 160 s of irradiation (just prior to transport) with UV light in the position shown (the right edge of the droplet and the adjacent surface). **g** After 245 s of irradiation (just after transport). **h** After 640 s irradiation (at the photostationary state). For clarity, on photographs **f–h** a *yellow line* is used to indicate the surface of the substrate

◀ **Scheme 18** A photo-responsive surface based on switchable fluorinated molecular shuttles. Light-switchable rotaxanes with the fluoroalkane region (*orange*) exposed (*E-37*) were physisorbed onto a SAM of 11-MUA on Au(111) deposited onto either glass or mica to create a polarophobic surface, E-37·11-MUA·Au(111). Illumination with 240–400 nm light isomerizes some of the E olefins to Z causing a nanometer displacement of the rotaxane threads in the Z-shuttles which encapsulates the fluoroalkane units leaving a more polarophilic surface, E/Z-37·11-MUA·Au(111). The contact angles of droplets of a wide range of liquids change in response to the isomerization process

5
Conclusions

The relative positioning of the components of hydrogen bonded benzylic amide catenanes and rotaxanes can be switched, rotated, speeded up, slowed down and directionally driven in response to a remarkable range of stimuli. In doing so they can affect the nanoscopic and macroscopic properties of the system to which they belong. Whether one chooses to call these and similar systems "motors" and "machines", or rather consider them more classically in terms of specific triggered large amplitude conformational, configurational and structural changes, to appreciate the overriding importance and potential of controlled molecular motion one only has to realize that it is at the heart of virtually every biological process. In contrast, at the start of the 21st century none of mankind's technology (with the notable exception of liquid crystals) exploits controlled molecular-level motion in any way at all. When we learn how to produce molecules that can bias random dynamic processes in response to stimuli in a controlled manner—and discover how to interface their effects with other molecules and the outside world —it will add a completely new dimension to functional molecule and materials design. An improved understanding of physics and biology will also surely follow. There is no question that much of this revolution lies far in the future but it is inevitable nonetheless. The first small steps along this path have been taken.

References

1. Balzani V, Credi A, Raymo FM, Stoddart JF (2000) Angew Chem Int Ed 39:3349
2. Balzani V, Venturi M, Credi A (2003) Molecular devices and machines. A journey into the nanoworld. Wiley-VCH, Weinheim
3. Easton CJ, Lincoln SF, Barr L, Onagi H (2004) Chem Eur J 10:3120
4. Kottas GS, Clarke LI, Horinek D, Michl J (2005) Chem Rev 105:1281
5. Kinbara K, Aida T (2005) Chem Rev 105:1377
6. Kay ER, Leigh DA (2005) Synthetic molecular machines. In: Schrader T, Hamilton AD (eds) Functional artificial receptors. Wiley-VCH, Weinheim, p 333
7. Astumian RD, Hänggi P (2002) Phys Today 55(11):33
8. Jones RAL (2004) Soft machines: nanotechnology and life. OUP, Oxford
9. Hänggi P, Bartussek R (1996) Lect Notes Phys 476:294
10. Jülicher F, Ajdari A, Prost J (1997) Rev Mod Phys 69:1269
11. Reimann P (2002) Phys Rep 361:57
12. Reiman P, Hänggi (2002) Appl Phys A 75:169
13. Parrondo JMR, de Cisneros BJ (2002) Appl Phys A 75:179
14. Gabryœ BJ, Pesz K, Bartkiewicz SJ (2004) Physica A 336:112
15. Astumian RD (1997) Science 276:917
16. Astumian RD (2002) Appl Phys A 75:193
17. Oster G, Wang H (2003) 13:114
18. Kurzyński M, Chełminiak P (2004) Physica A 336:123

19. Schliwa M (ed) (2003) Molecular motors. Wiley-VCH, Weinheim
20. Hess H, Vogel V (2001) Rev Mol Biotechnol 82:67
21. Hess H, Bachand GD, Vogel V (2004) Chem Eur J 10:2110
22. Purcell EM (1977) Am J Phys 45:3
23. Schill G (1971) Catenanes, rotaxanes and knots. Academic Press, New York
24. Walba DM (1985) Tetrahedron 41:3161
25. Amabilino DB, Stoddart JF (1995) Chem Rev 95:2725
26. Leigh DA, Murphy A (1999) Chem Ind 178
27. Breault GA, Hunter CA, Mayers PC (1999) Tetrahedron 55:5265
28. Sauvage J-P, Dietrich-Buchecker CO (eds) (1999) Molecular catenanes, rotaxanes and knots. Wiley-VCH, Weinheim
29. Hannam JS, Kidd TJ, Leigh DA, Wilson AJ (2003) Org Lett 5:1907
30. Block SM (1996) Cell 87:151
31. Block SM (1998) Cell 93:5
32. Skou JC (1998) Angew Chem Int Ed 37:2321
33. Fyfe MCT, Glink PT, Menzer S, Stoddart JF, White AJP, Williams DJ (1997) Angew Chem Int Ed Engl 36:2068
34. Wasserman E (1960) J Am Chem Soc 82:4433
35. Schill G, Lüttringhaus A (1964) Angew Chem Int Ed Engl 3:546
36. Harrison IT, Harrison S (1967) J Am Chem Soc 89:5723
37. Harrison IT (1974) J Chem Soc Perkin Trans 1:301
38. Dietrich-Buchecker CO, Sauvage J-P, Kintzinger JP (1983) Tetrahedron Lett 24:5095
39. Dietrich-Buchecker CO, Sauvage J-P (1987) Chem Rev 87:795
40. Sauvage J-P (1990) Acc Chem Res 32:53
41. Busch DH, Stephenson NA (1990) Coord Chem Rev 100:119
42. Hoss R, Vögtle F (1994) Angew Chem Int Ed Engl 33:375
43. Philp D, Stoddart JF (1996) Angew Chem Int Ed Engl 35:1155
44. Fujita M, Ogura K (1996) Coord Chem Rev 148:249
45. Fyfe MCT, Stoddart JF (1997) Acc Chem Res 30:393
46. Jager R, Vögtle F (1997) Angew Chem Int Ed Engl 36:930
47. Diederich F, Strang PJ (eds) (1999) Templated organic synthesis. Wiley-VCH, Weinheim
48. Fujita M (1999) Acc Chem Res 32:53
49. Hubin TJ, Busch DH (2000) Coord Chem Rev 200:5
50. Furlan RLE, Otto S, Sanders JKM (2002) Proc Natl Acad Sci USA 99:4801
51. Chambron J-C, Collin J-P, Heitz V, Jouvenot D, Kern J-M, Mobian P, Pomeranc D, Sauvage J-P (2004) Eur J Org Chem 1627
52. Hunter CA (1992) J Am Chem Soc 114:5303
53. Vögtle F, Meier S, Hoss R (1992) Angew Chem Int Ed Engl 31:1619
54. Johnston AG, Leigh DA, Pritchard RJ, Deegan MD (1995) Angew Chem Int Ed Engl 34:1209
55. Johnston AG, Leigh DA, Murphy A, Smart JP, Deegan MD (1996) J Am Chem Soc 118:10662
56. Johnston AG, Leigh DA, Nezhat L, Smart JP, Deegan MD (1995) Angew Chem Int Ed Engl 34:1212
57. Leigh DA, Venturini A, Wilson AJ, Wong JKY, Zerbetto F (2004) Chem Eur J 10:4960
58. Leigh DA, Murphy A, Smart JP, Slawin AMZ (1997) Angew Chem Int Ed Engl 36:728
59. Lane AS, Leigh DA, Murphy A (1997) J Am Chem Soc 119:11092
60. Gatti FG, Leigh DA, Nepogodiev SA, Slawin AMZ, Teat SJ, Wong JKY (2001) J Am Chem Soc 123:5983

61. Leigh DA, Murphy A, Smart JP, Deleuze MS, Zerbetto F (1998) J Am Chem Soc 120:6458
62. Deleuze MS, Leigh DA, Zerbetto F (1999) J Am Chem Soc 121:2364
63. Leigh DA, Troisi A, Zerbetto F (2001) Chem Eur J 7:1450
64. Sandström J (1982) Dynamic NMR spectroscopy. Academic Press, London
65. Dahlquist FW, Longmur KJ, Du Vernet RB (1975) J Magn Reson 17:406
66. Leigh DA, Troisi A, Zerbetto F (2000) Angew Chem Int Ed 39:350
67. Anelli PL, Spencer N, Stoddart JF (1991) J Am Chem Soc 113:5131
68. Keaveney CM, Leigh DA (2004) Angew Chem Int Ed 43:1222
69. Brouwer AM, Frochot C, Gatti FG, Leigh DA, Mottier L, Paolucci F, Roffia S, Wurpel GWH (2001) Science 291:2124
70. Altieri A, Gatti FG, Kay ER, Leigh DA, Martel D, Paolucci F, Slawin AMZ, Wong JKY (2003) J Am Chem Soc 125:8644
71. Leigh DA, Pérez EM (2004) Chem Commun 2262
72. Altieri A, Bottari G, Dehez F, Leigh DA, Wong JKY, Zerbetto F (2003) Angew Chem Int Ed 42:2296
73. Bottari G, Dehez F, Leigh DA, Nash PJ, Pérez EM, Wong JKY, Zerbetto F (2003) Angew Chem Int Ed 42:5886
74. Hanke A, Metzler R (2002) Chem Phys Lett 359:22
75. Bermudez V, Capron N, Gase T, Gatti FG, Kajzar F, Leigh DA, Zerbetto F, Zhang SW (2000) Nature 406:608
76. Gatti FG, León S, Wong JKY, Bottari G, Altieri A, Morales MAF, Teat SJ, Frochot C, Leigh DA, Brouwer AM, Zerbetto F (2003) Proc Natl Acad Sci USA 100:10
77. Lukin O, Kubota T, Okamoto Y, Schelhase F, Yoneva A, Müller WM, Müller U, Vögtle F (2003) Angew Chem Int Ed 42:4542
78. Leigh DA, Moody K, Smart JP, Watson KJ, Slawin AMZ (1996) Angew Chem Int Ed Engl 35:306
79. Ceroni P, Leigh DA, Mottier L, Paolucci F, Roffia S, Tetard D, Zerbetto F (1999) J Phys Chem B 103:10171
80. Leigh DA, Wong JKY, Dehez F, Zerbetto F (2003) Nature 424:174
81. Kelly TR, De Silva H, Silva RA (1999) Nature 401:150
82. Kelly TR (2001) Acc Chem Res 34:514
83. Sestelo JP, Kelly TR (2002) Appl Phys A 75:337
84. Koumura N, Zijlstra RWJ, van Delden RA, Harada N, Feringa BL (1999) Nature 401:152
85. Kelly TR, Silva RA, De Silva H, Jasmin S, Zhao YJ (2000) J Am Chem Soc 122:6935
86. Feringa BL (2001) Acc Chem Res 34:504
87. Feringa BL, Koumura N, van Delden RA, ter Wiel MKJ (2002) Appl Phys A 75:301
88. Koumura N, Geertsema EM, van Gelder MB, Meetsma A, Feringa BL (2002) J Am Chem Soc 124:5037
89. Geertsema EM, Koumura N, ter Wiel MKJ, Meetsma A, Feringa BL (2002) Chem Commun 2962
90. ter Wiel MKJ, van Delden RA, Meetsma A, Feringa BL (2003) J Am Chem Soc 125:15076
91. Feringa BL, van Delden RA, ter Wiel MKJ (2003) Pure Appl Chem 75:563
92. Hernández JV, Kay ER, Leigh DA (2004) Science 306:1532
93. Maxwell JC (1867) Letter to PG Tait, 11 December 1867. Quoted in: Knot CG (1911) Life and scientific work of Peter Guthrie Tait. Cambridge University Press, Cambridge, p 213

94. Pérez EM, Dryden DTF, Leigh DA, Teobaldi G, Zerbetto F (2004) J Am Chem Soc 126:12210
95. Asakawa M, Brancato G, Fanti M, Leigh DA, Shimizu T, Slawin AMZ, Wong JKY, Zerbetto F, Zhang SW (2002) J Am Chem Soc 124:2939
96. Bottari G, Leigh DA, Pérez EM (2003) J Am Chem Soc 125:13360
97. Leigh DA, Morales MAF, Pérez EM, Wong JKY, Saiz CG, Slawin AMZ, Carmichael AJ, Haddleton DM, Brouwer AM, Buma WJ, Wurpel GWH, León S, Zerbetto F (2005) Angew Chem Int Ed 44:3062
98. Berná J, Leigh DA, Lubomska M, Mendoza SM, Pérez EM, Rudolf P, Teobaldi G, Zerbetto F (2005) Nature Mater 4:704

Amphidynamic Crystals: Structural Blueprints for Molecular Machines

Steven D. Karlen · Miguel A. Garcia-Garibay (✉)

Department of Chemistry and Biochemistry, University of California, Los Angeles, California, CA 90095-1569, USA
mgg@chem.ucla.edu

1	**Introduction**	180
1.1	Molecular Machines	182
1.2	Crystalline Molecular Machines	183
1.3	Amphidynamic Crystals: Molecular Blueprints for Rapid Dynamics in Close-Packed Systems	183
1.4	Emulating Machines and Machine Parts with Close-Packed Molecular Systems	185
2	**Gyroscopes**	187
2.1	Molecular Gyroscopes and Molecular Compasses	187
2.2	Molecular Blueprints	188
3	**Determination of Gyroscopic Dynamics in the Solid-State by Nuclear Magnetic Resonance**	190
3.1	Cross Polarization with Magic Angle Spinning ^{13}C NMR and Dipolar Dephasing (^{13}C CPMAS-DD)	190
3.2	^{13}C NMR Coalescence Studies Using CPMAS	191
3.3	Lineshape Analysis of Quadrupolar Echo ^{2}H NMR	192
3.4	Spin-Lattice Relaxation	194
4	**Molecular Gyroscopes with Triptycyl Stators**	195
4.1	A Molecular Gyroscope with a bis-Triptycyl Stator and a Phenylene Rotator	198
4.2	Towards Structural Refinements in Triptycyl Stators	200
5	**Molecular Gyroscopes with Triphenylmethyl Stators**	203
5.1	A Molecular Gyroscope with a bis-Triphenylmethyl Stator and a Phenylene Rotator	204
5.2	Structural Diversity Through Polymorphism in Molecular Gyroscopes with Substituted Trityl Stators	209
5.3	Steric Shielding Strategies in Molecular Gyroscopes with Trityl Stators	213
6	**Rotator Effects**	215
6.1	Arylene Rotators	216
6.2	Phenylene Rotators with Polar Groups: Molecular Compasses	217
6.3	Dual Rotators: Phenylene—Diamantane	220
7	**Conclusions and Outlook**	224
	References	225

Abstract By considering the relation between molecular structure, molecular dynamics, and phase order, we suggest that certain structures should be able to make up supramolecular assemblies with structurally programmed molecular dynamics. Given that the simplest members of these structures should have the elements required to form a rigid lattice with moving parts, we propose the term amphidynamic crystals to describe them. We also suggest that amphidynamic crystals may form the basis of a new class of molecular machines with a function that is based on the collective dynamics of molecules within crystals. These crystalline molecular machines represent an exciting new branch of crystal engineering and materials science. In this chapter, we described current efforts for the construction of crystalline molecular machines designed to respond collectively to mechanic, electric, magnetic, and photonic stimuli, in order to fulfill specific functions. One of the main challenges in their design and construction derives from the picometric precision required for operation within the close-packed, self-assembled environment of crystalline solids. We illustrate here some of the initial advancements in the preparation, crystallization, and dynamic characterization of these interesting systems.

Keywords Crystal engineering · Dielectric spectroscopy · Dynamics in crystals · Molecular machines · Solid state NMR · X-ray diffraction

1
Introduction

Macroscopic machines tend to be densely packed, complex assemblies, with many components acting together in a synchronized manner to transfer motion, force, or energy from one to another [1]. Some machines are designed to reduce the amount of work required to accomplish a given task, and some are designed to acquire, process, and/or store information. While their function and design may vary considerably, the construction of macroscopic machines is based on information that describes precisely their structure and dynamics. Using a wristwatch as an example, one would start by drafting the blueprints that describe the required parts, including wheels, ratchets, pivots, shafts, screws, cocks, barrels, and many other components, each with the proper materials. When assembled correctly and given a proper input, about 200 components come together to form a properly functioning mechanical watch (Fig. 1) [2]. Each of the non-spherical N-rigid macroscopic components has 3 translational and up to 3 rotational degrees of freedom. When assembled, as many as 6N-7 degrees of freedom are lost in the process with the watch as a whole retaining only 6 external degrees of freedom and as few as 1 internal degree of freedom which defines the function of the assembly [3].

With an intuition that arises from handling macroscopic molecular models, chemists have recognized for a long time that force, motion, and energy can be transferred within molecules, or from one molecule to another, in a manner that depends on the structure of the assembly and its medium [4, 5]. With atoms assembled by chemical (covalent and non-covalent) bonds, one may speculate that the internal degrees of freedom of a molecular ensem-

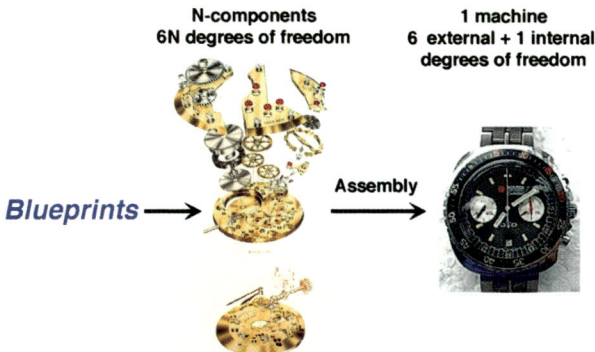

Fig. 1 The blueprints contain the structural information necessary to build and assemble the components of a mechanical wristwatch. Each component must have the precise structural information to fit within the assembly and to surrender its own degrees of freedom to the degrees of freedom of the machine [2]

ble can be programmed into the molecular blueprints to carry out desirable mechanical processes. However, to embrace the construction of molecular machines as envisioned by Feynman over 40 years ago [6], chemists had to wait for sound structural theories, more powerful synthetic methods,

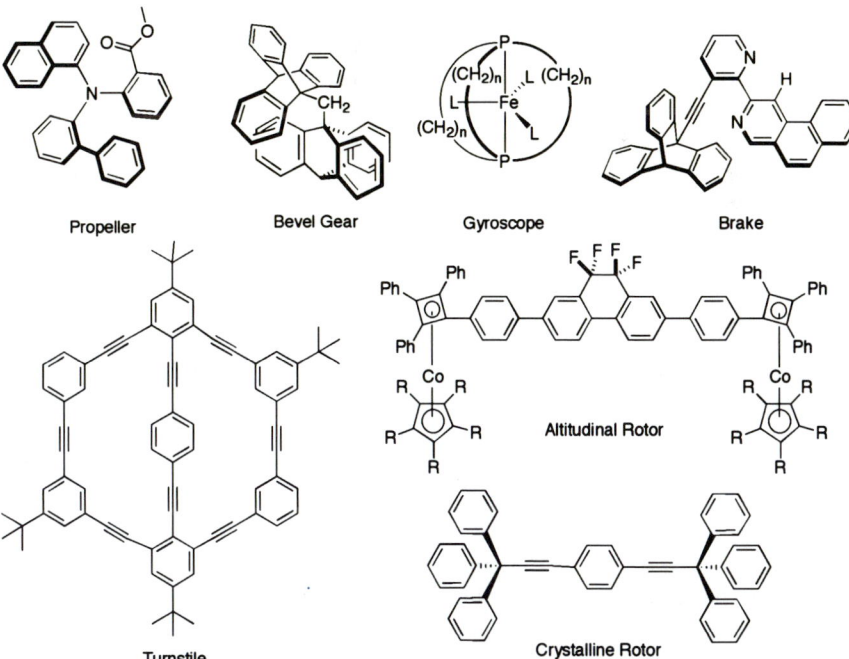

Fig. 2 Molecular analogs to macroscopic objects based on rotors [2, 10–16]

and reasonably good tools to determine molecular dynamics. In addition, conceptual barriers on how to approach the creation of artificial molecular machines are slowly coming down as a result of a greater understanding of structure-property relations in biological motors [7, 8] and the intuitive appreciation that comes from methods to determine the dynamics of single molecules [9]. While work on artificial "molecular machines" remains in a conceptual stage, by designing and synthesizing crude analogical models intended to emulate the dynamics of macroscopic objects, it is expected that progress in the field will occur at an accelerated pace over the next few years. Early examples involving the construction of molecular propellers [10], gears [11], brakes [12], turnstiles [13], shuttles [14], gyroscopes [2,15], surface mounted rotors [16], and of many other "objects" [17], have roused the imagination of a growing community, and have helped expose some of the challenges involved in the design and construction of truly serviceable molecular machines (Fig. 2).

1.1
Molecular Machines

Naturally, there are many distinctions between macroscopic and molecular machines [6, 7]. While the internal motion and mechanical functions of macroscopic machines may be described well by Newtonian dynamics and deterministic models, the internal motion of molecular machines is subject to dissipation (thermal) forces and distributions of trajectories that can only be described in a probabilistic manner [7, 18, 19]. Macroscopic machine parts can be welded together into rigid bodies with no internal degrees of freedom, but every additional atom or group of atoms in a molecular machine continues adding to the 3N-6 internal degrees of freedom of the system. Most macroscopic machines have continuous velocities within a given range and have absolute resting states when turned off. In contrast, the transfer of motion and thermal energy redistribution within molecular machines occurs between quantized states, with motion never stopping, not even at absolute zero. While the function of macroscopic machines is based on the use of the inertia acquired by the mobile parts of the system, molecular machines are strongly overdamped and dissipate their momenta very rapidly. Using an anthropomorphic analogy and considering the magnitude of forces involved under conditions where molecular machines are expected to function, Astumian pointed out that in order to appreciate the function of biomolecular machines one should imagine what it would be like to "swim in molasses and walk in hurricanes". He notes that the function of biomolecular machines is based on mechanisms where molecular motion is modified by suitable (mostly chemical) inputs that change the relative energetics of the machine's equilibrium and transition states [20].

1.2
Crystalline Molecular Machines

While much of the early research on molecular machines addresses the synthesis, characterization, and eventual manipulation of isolated molecular system in solution, we have decided to center our efforts on supramolecular assemblies. The thesis behind our work rests on the premise that information contained at the molecular level in the form of composition, topology, size, shape, non-bonding interactions, electronic structure, etc., will dictate the aggregation, dynamics, switching, and potential mechanical functions [21]. One of our goals is to uncover the relation between supramolecular structure and intermolecular dynamics in search of lessons that may lead to functional molecular systems [2]. In order to document the structure-dynamics relation with as high precision and certainty as possible, and knowing that machines tend to be dense multicomponent objects, we have chosen the crystalline solid state as a promising model. Structural characterizations can be carried out by single crystal X-ray diffraction measurements, and dynamic properties can be determined over a wide range of time scales by techniques that include variable temperature solid state nuclear magnetic resonance (NMR), dielectric spectroscopy, and inelastic neutron scattering, among several others.

1.3
Amphidynamic Crystals: Molecular Blueprints for Rapid Dynamics in Close-Packed Systems

Molecular crystals are highly ordered three-dimensional arrays held together by relatively weak non-bonding interactions [22]. Crystals are characterized by their periodicity, homogeneity, and unique collective properties. As molecular crystals are assembled into close packing arrangements that minimize the amount of empty space [23], molecules are forced to surrender their individual degrees of freedom in favor of a new set of degrees of freedom, which are characteristic of the ensemble. While thermal energy in solution and the gas phase is distributed between vibrational, rotational, and translational degrees of freedom, the thermal energy of molecules in crystals is distributed over its normal modes, oscillations with small angular displacements or librations of groups linked by single bonds, and the collective small amplitude displacements of molecules in the lattice, or phonons, which may be visualized as ocean waves [24].

In search of an insight into how structure influences dynamics one may analyze the relationship between molecular motion and long-range order as illustrated in Fig. 3. With a measure of order on the vertical axis and a measure of motion on the horizontal axis, we locate liquids in the bottom right corner of the plot. Liquids have very low long-range order and each molecule exhibits rapid full body rotation, molecular translation in all

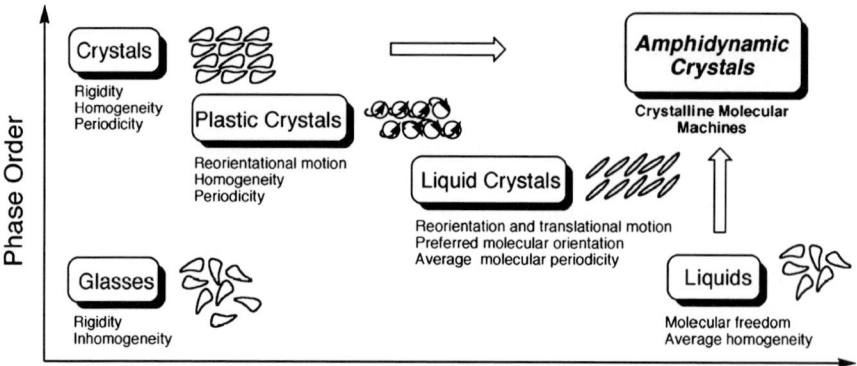

Fig. 3 Phase order-molecular motion diagram illustrating possible forms of condensed phase matter. Crystalline molecular machines would require a combination of rigidity and mobility to warrant a high level of order and desirable dynamic processes [2]

three directions, and frequent collisions with continuously exchanging neighboring molecules. The properties of liquids are isotropic, which means that they are identical in all directions, by virtue of their rapid molecular dynamics, which renders them homogeneous on a time average. At the other extreme, located at the top left corner of the plot, are molecular crystals of arbitrarily shaped molecules. Molecular crystals are characterized by their rigidity, homogeneity, periodicity, and their macroscopic anisotropy. Molecular crystals have long-range order and only the small-amplitude molecular displacements associated with lattice vibrations. Molecules in crystals retain their lattice positions for very long periods of time and most physical properties vary along the different crystallographic directions in a manner that depends on the lattice symmetry. Near the origin of the graph, with no long-range order and minimal molecular motion (below their glass transition temperatures) are amorphous glasses [25]. As a snapshot of a liquid, glasses are highly inhomogeneous with many molecular conformations and intermolecular arrangements. The physical properties of macroscopic glasses, like those of liquids, are isotropic. Filling in the gap between the three corners are the mesophases formed by liquid crystals and plastic crystals. Molecules that form plastic crystals are characterized by having high-symmetry structures with spherical or cylindrical shapes [26, 27]. While their center of mass remains fixed on a given lattice position, the entire molecule experiences very fast rotational motion, either continuously or by jumping between equivalent sites. Examples of plastic crystal-forming molecules are adamantane, neopentane, and many long chain alkanes [28]. Next in the plot, with an increase in molecular motion and a decrease in molecular order are the liquid crystals [26, 29]. With significantly more translational, rotational and conformational freedom, liquid crystals consist of molecules with rigid, rod- and disk-shaped cores, attached to two or more floppy chains. Molecules in liquid crystals retain preferred molecular orientations and average long-range molecular order, yet they experience fast rotation and diffusion.

The fact that molecular structure determines the formation and properties of plastic and liquid crystals, as well as their dynamics, suggests that suitable molecular structures may be designed to form a highly ordered rigid lattice capable of supporting rapid motion of well-designed components. One can imagine structures built with rigid frames, axles, pivots, rotary elements, and oscillating components may be the basis of a new class of crystalline molecular machines. We suggest that condensed phase media above the top-left to bottom-right diagonal should provide many interesting opportunities in materials science. On the top right corner, one may envision crystals where portions of the assembly can be constrained to specific lattice positions while other components exhibit rapid molecular motion. As suggested by the presence of a rigid ordered component and a highly mobile component, we would like to define these phases as "amphidynamic" solids, where the suffix "amphi" stands for "both sides" [2].

1.4
Emulating Machines and Machine Parts with Close-Packed Molecular Systems

Mechanical motions at the molecular level require two or more components interacting with each other within the constraints generated by their potential energy surfaces. The simplest mechanical processes within molecules are internal vibrations, such as bond stretching and bond deformation, or bending [30], which also serve as the primary depository for the internal kinetic energy of molecules. Internal rotations derive from thermal excitation of rocking and torsion deformations [31], and may be viewed as the residual rotational degrees of freedom of a group of atoms relative to the frame of reference of another group of atoms to which it is linked by a single bond. Complex mechanical motions derive from the population of higher vibrational levels that increase the kinetic energy of the system to help overcome internal barriers and allow the structure to explore larger regions of its potential energy surface. To design mechanical motions within molecules will require a good understanding of how structure determines the potential energy surface of the system. Complex dynamics may include correlated and synchronous rotation about several bonds [32], association-dissociation in host-guest complexes [33], the relative displacement of the components that make up catenanes and rotaxanes [34], and more complex processes in multicomponent self-assemblies and nanostructures [35, 36].

In order to engineer rapid dynamics within crystalline solids one must have a good understanding of molecular and crystal potentials and a good insight into the coupling of molecular vibrations and lattice phonons. As a starting point to begin a systematic investigation of mechanical processes in crystals one may emulate the types of structures that are known to work in macroscopic machines and biomolecular systems. For example, to perform complex functions with a high density of moving parts, mechanical clocks, automobile engines, and mechanical typewriters must be structurally programmed with a very careful design. A large number of interacting components must fulfill precise metric and symmetry relationships that allow them to maintain their motions correlated. Their internal dynamics are based on volume-conserving motions, which tend to be periodic, either rotatory or oscillatory. To be tolerated in a close-packed, highly dense environment, these motions must either have a small amount of built-in free volume, or must be precisely complementary and highly correlated.

The molecular blueprints for the construction of amphidynamic crystals should include components that function as static frames, or stators, designed to support the dynamic portions of the system. Highly mobile components may be isolated from each other, or may be designed with strong steric or electrostatic interactions that propagate their state of motion over long distances within the crystal lattice. The stator should connect to the moving components through suitable axles that support their rotary motion, or through pivots that support an oscillatory movement. Using stators, rotators, and axles as the key structural elements, we recognized that macroscopic gyroscopes provide one of the most promising designs to be emulated at the molecular level for the realization of amphidynamic crystals (Fig. 4).

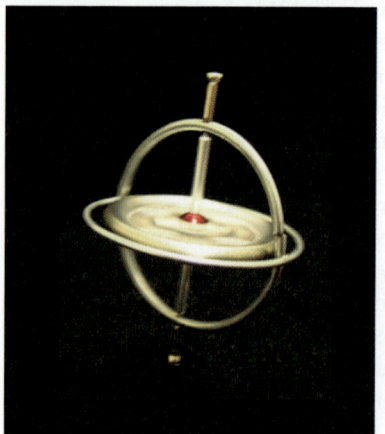 

Fig. 4 A macroscopic gyroscope (*left*) and a macroscopic model of a lattice made up of gyroscopes (*right*)

2
Gyroscopes

Macroscopic gyroscopes are used as navigational devices to sense changes in orientation regardless of their frame of reference by a function that relies on the conservation of angular momentum. A simple gyroscope is a device consisting of a spinning mass, or rotator, with a spinning axis that projects through the center of the mass, which is mounted within a rigid frame, or stator [37]. Simple gyroscopes rely on a rigid frame and one internal degree of freedom. Complex gyroscopes used in the inertial navigation systems (INS) of airplanes and satellites are mounted on gimbals in order to have additional degrees of freedom, while their state of motion is preserved by a power source.

A single toy gyroscope is illustrated in Fig. 4 along with a model amphidynamic crystal built with a collection of macroscopic gyroscopes. With proper encasing frames, gyroscopes may be designed to sustain a state of motion when either isolated or as an ensemble. The characteristic physical properties of both isolated units and aggregates are dependent on their state of motion. The angular momentum of each gyroscope is defined by the direction of its rotation axis and is proportional to its angular velocity. The total angular momentum of the ensemble would be the sum of the angular momenta of all of its components. While collections of randomly oriented gyroscopes would have no angular momentum, maximum values would be expected when all the units in the ensemble have their rotational axis collinear and rotating in the same direction with respect to a common reference frame. Such an ideal gyroscope crystal would resist changes in orientation by exchanging its internal energy with acting forces that cause its reorientation and rotation.

2.1
Molecular Gyroscopes and Molecular Compasses

It is not obvious how the physics of a macroscopic gyroscope, and their hypothetical crystals, would scale from the macroscopic to the molecular levels. Among the differences expected are the previously mentioned overdamping of motion, when frictionless rotation or a power source to maintain a state of motion would be required. Furthermore, for a crystal of molecular gyroscopes to emulate the properties of the macroscopic model, one would have to control molecular alignment and the directionality of rotation within all the molecules in the ensemble. While it is unlikely that molecular gyroscopes will act as their macroscopic namesakes, the potential of controlling rotary dynamics in oriented molecular systems is a fascinating scientific challenge in its own right. In fact, given the large number of physical properties affected by internal motion, many potential material science applications may be envisioned. For example, by endowing the rotary element of a molecular gyroscope with a dipole moment (Fig. 5), its structure may emulate the function

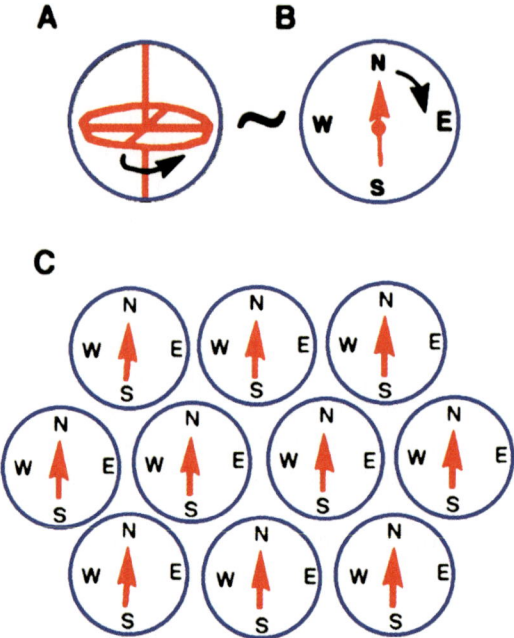

Fig. 5 Topological analogies between the structures of a gyroscope (**A**) and a compass (**B**) and a hypothetical crystal of molecular compasses (**C**) showing a ground state ferroelectric alignment

of a macroscopic compass. Applications for crystals of molecular compasses may be devised with functions that rely on the orientation-dependent energy of their internal dipoles with respect to external fields. Molecular compasses may be envisioned with electric or magnetic dipoles so that their bulk properties may be affected by external electric, magnetic, and electromagnetic fields. In fact, many interesting photonic, dielectric, and phonon phenomena are expected from arrays of molecular compasses. These include addressable ferroelectric and anti-ferroelectric phases, polar rotary phonons that propagate at velocities slower than the typical speed of sound, and Nambu–Goldstone dipole-wave quanta [38]. One of the most interesting aspects of long-range crystalline arrangements is that the absolute direction of molecular motion can be controlled with respect to macroscopic frames of reference simply by rotating the crystal along previously indexed crystallographic directions.

2.2
Molecular Blueprints

The first step towards the realization of a macroscopic object at the molecular level is to select the atomic and molecular components that may emulate the desired structure and its function. In the case of molecular gyroscopes, systematic structural variations should allow for a determination of a structure-activity relationship to correlate the effects of molecular structure on crystal packing, and the effect of both molecular and packing structures on molecular dynamics. As illustrated in Fig. 6, the key elements for the construction of a molecular gyroscope are a rotator, a stator, and an axle that joins them. The rotator may be any axially symmetric group with its center of mass aligned along a single bond that supplies both the rotary axis and the point of attachment to the static framework. Potentially interesting rotators may be molecular fragments such as 1,4-phenylenes, 1,4-bicyclo[2.2.2]octanes, diamantanes, carboranes, etc. These are represented as a simple sphere in Fig. 6. The axle of rotation for an ideal gyroscope should have no intrinsic rotational barrier, so that the rate of motion may be controlled over a wide dynamic range with built-in steric interactions in the stator, or through chemical changes that modify one of three components. In this chapter, we will illustrate the use of two collinear acetylene groups for the rotary axle. Not only are alkynes exceedingly versatile from a synthetic perspective, but the intrinsic barriers for rotation about single bonds formed between sp-sp^n carbons are ca. 0.1 kcal/mol or less, as suggested both by semiempirical quantum mechanical calculations and experimental observations using microwave spectroscopy [39]. Finally, the stator should provide an encapsulating frame to shield the rotator from steric contacts with adjacent molecules in the crystal. The ideal stators should completely encapsulate the rotator and guide the formation of suitable crystal structures. The structure of the stator should also provide a tunable rotational barrier and externally addressable elements that may help change the equilibrium dynamics of the central rotator. Some of the simplest and most promising stators with and without fully encapsulating structures can be built with triptycenes and triphenyl methanes as illustrated in Fig. 6.

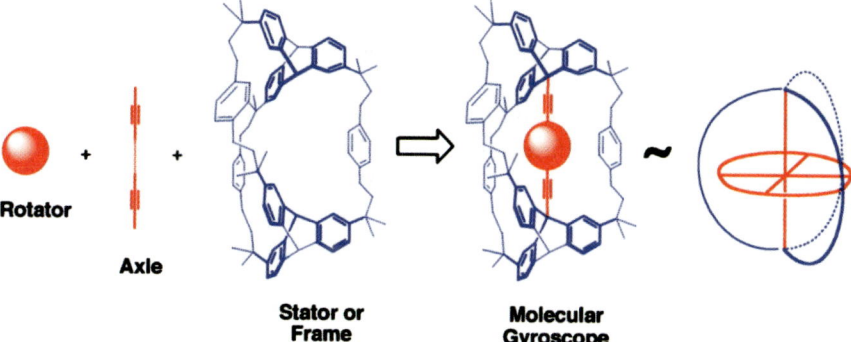

Fig. 6 Suggested components for the realization of functional molecular gyroscopes to build amphidynamic crystals. The effects of various stator and rotator structures on the dynamics of the crystalline solid is discussed in Sects. 4–6

3
Determination of Gyroscopic Dynamics in the Solid-State by Nuclear Magnetic Resonance

The study of molecular motions by NMR has become a standard practice in modern chemistry. While many 1D and 2D techniques have been developed to investigate the dynamics of both solution and solid samples by taking advantage of the time-dependence of a wide range of magnetic interactions, we will only discuss here some of the simplest and most accessible methods used in our own research in the last few years [40]. First we will describe the use of high resolution solid state NMR by cross polarization and magic angle spinning (CPMAS) followed by a brief discussion of quadrupolar echo ^{2}H NMR with static samples, and then the use of spin-lattice relaxation (T_1) measured under CPMAS conditions.

3.1
Cross Polarization with Magic Angle Spinning ^{13}C NMR and Dipolar Dephasing (^{13}C CPMAS-DD)

Solution-like spectra can be obtained with solid samples by taking advantage of the cross polarization and magic angle spinning [41, 42]. The term cross polarization (CP) relates to the transfer of magnetization from the abundant and sensitive ^{1}H to the less sensitive and dilute nuclei, such as ^{13}C. The experimental procedure includes simultaneous high power ^{1}H-decoupling and fast sample spinning (5–20 kHz) at the magic angle (54.7°) to remove line broadening from anisotropic interactions mediated by the external magnetic field, such as heteronuclear dipolar coupling and chemical shift anisotropy [41]. Although resolution and line widths in the solid state are not as good as in solution, chemical shifts determined by the CPMAS method are generally comparable. ^{13}C CPMAS is a valuable analytical tool to document the phase purity of solid samples, the occurrence of phase transitions, and the number of crystallographic and magnetically non-equivalent signals for a given crystal structure.

One of the simplest ways to document molecular motion in the solid state is by using a simple modification of the CPMAS experiment known as dipolar dephasing (DD), also known as non-quaternary signal suppression (NQS) [43]. The CPMAS-DD experiment consists of turning off the high power ^{1}H-decoupler immediately after cross polarization but before signal acquisition. In this experiment, a strong C–H dipole-dipole interaction of ca. 20 kHz for static C–H groups dephases the carbon polarization, so that the corresponding signal disappears from the spectrum. However, the strength of the dipole-dipole interaction of C–H groups that are part of highly mobile substituents or molecules can be diminished considerably by dynamic averaging, resulting in the persistence of the corresponding signals after the DD pulses. Thus, the solid state ^{13}C CPMAS-DD signals of highly mobile methyl groups and rapidly rotating molecules such as adamantane can be detected, even after relatively long decoupling periods (ca. 50 µs). It is expected that highly dynamic rotators will present signals that remain after the CPMAS-DD sequence while the signals corresponding to the stator should disappear by virtue of their strong dipole-dipole interactions [43].

3.2
^{13}C NMR Coalescence Studies Using CPMAS

Rotators with atoms occupying two or more magnetically non-equivalent positions may experience a dynamic process where these atoms jump from site to site. The rate of exchange (k, in Hz) between sites is governed by the energy barriers along the rotational potential and by the sample temperature. Variations in the spectrum as a function of temperature provide information on the exchange rate within certain dynamic windows (Fig. 7) [44]. There are three exchange regimes, slow, intermediate, and fast, which are defined with respect to the so-called NMR time scale. The NMR time scale for atom exchange is determined by the difference in chemical shift (Δv, in Hz) between dynamically related sites. When the rate of exchange is fast ($k \ll \Delta v$) the signals of the two sites are averaged into one chemical shift, which is seen as a sharp signal. When the motion is slow ($k \gg \Delta v$) the symmetry equivalent sites retain their identify during the NMR time scale and appear as distinct, sharp signals. In the intermediate exchange case, the rate of site exchange in Hz is close to the difference in chemical shift ($k \approx \Delta v$) and the two signals collapse into one very broad peak, in a process that is referred to as coalescence.

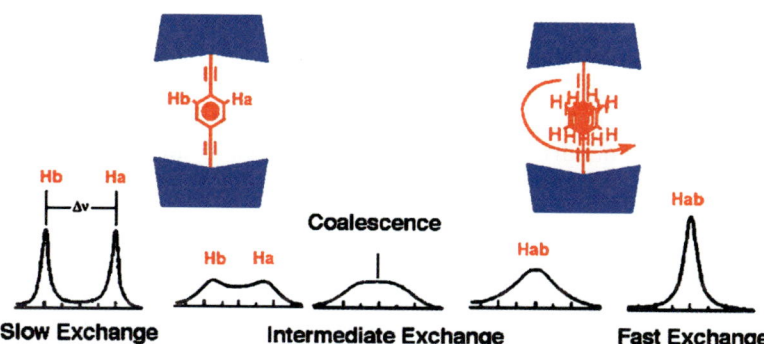

Fig. 7 Changes in the spectrum of a hypothetical molecular gyroscope with a phenylene rotator as a function of rotational correlation time showing the coalescence of two signals Ha and Hb separated by Δv Hz as they go from the slow, to the intermediate, to the fast exchange regime. The correlation time at the coalescence temperature (T_c) is given by: $\tau_c = 2^{1/2}/(\pi \Delta v)$

The rate of exchange k_c at the temperature at which the signals coalesce (T_c) is given by $k_c = \pi \Delta v (2)^{-1/2} \approx 2.22 \Delta v$. As ambiguities are often encountered identifying the precise value of T_c, a more involved analysis is required to determine the precise exchange rates and the activation energy for the process. Ideally, experimental data should be acquired as a function of temperature over a range that goes from the slow to the fast exchange regimes so that changes in lineshape may be fit or simulated with a model that considers the position of each exchanging signal, the exchange trajectories, and the rate of exchange [45]. The rate data obtained by the model can be used to construct an Arrhenius plot, $\ln k$ vs. $1/T$, from which the pre-exponential factor (k_o) and the activation energy (Ea) for the exchange process can be determined, Eq. 1.

$$k = k_o \exp(-Ea/RT) \tag{1}$$

Given that ^{13}C CPMAS NMR line-widths in the solid state are typically 20 Hz or greater, the difference in chemical shift between exchanging nuclei should be ~ 50 Hz or more. The lowest exchange rates that can be measured by this method are resolution-limited and of the order of ca. 100 s^{-1}. Faster exchange rates may be determined for sites that have greater Δv values. An approximate upper limit for accurate measurements by this method is ca. 10^3 Hz, provided that the coalescing signals do not overlap with other signals in the spectrum.

3.3
Lineshape Analysis of Quadrupolar Echo ^{2}H NMR

Wide line deuterium NMR is a powerful technique for studying molecular order and molecular dynamics in the solid state [46, 47]. Deuterium (^{2}H) has a nuclear spin $I = 1$ and most of its NMR properties in the solid state are governed by quadrupolar interactions with the electric field gradient tensor. For example, a single crystal with only one type of C$-^2$H bond and with a symmetric electric field gradient tensor would have a spectrum consisting of a doublet with quadrupolar splitting (Δv) related to the orientation angle β of the C$-$H bond with respect to the external field.

$$\Delta v = 3/4(e^2 q_{zz} Q/h)(3\cos^2 g\beta - 1) = 3/4 QCC(3\cos^2 g\beta - 1) \tag{2}$$

Q in the equation represents the electric quadrupole moment of the deuteron, e and h are the electric charge and Plank constant, respectively, and q_{zz} is the magnitude of the principal component of electric field gradient tensor, which lies along the C$-^2$H bond. The QCC values depend on the type of ^{2}H and vary from ca. 150–200 KHz. The peak separation for static deuterons varies from ca. $\Delta v = 300$ KHz for $\beta = 0°$, to $\Delta v = 0$ when the C$-^2$H bond is at the magic angle $\beta = 54.7°$.

Deuterium is only 0.0156% abundant and its sensitivity is 0.00965 with respect to that of ^{1}H. While this makes the acquisition of natural abundance ^{2}H NMR spectra impractical, the use of selective isotopic labeling at only those positions of interest makes it a very sensitive probe for molecular dynamics in the solid state. Rather than using single crystals, it is more practical to use static powdered samples with crystallites having molecules oriented with all possible values of β. The corresponding spectrum consists of a collection of doublets from C–^{2}H bonds in all possible orientations, which give rise to a broad symmetric spectrum known as a Pake or powder pattern, characterized by two maxima and two shoulders (Fig. 8). Since the very large $\Delta\nu$ values observed in static samples require very short and powerful excitation pulses, the spectrum is measured with a spin-echo sequence. Systematic variations in the powder pattern occur when the C–^{2}H bonds experience dynamic exchange between sites that have different $\Delta\nu$ values. The internal dynamics of substituents and full body molecular reorientations result in the dynamic averaging of the magnetic interactions for different molecular orientations. Spectral lineshapes can be analyzed in terms of dynamic models that consider the types and rates of motion. Strong variations in the ^{2}H NMR spectra can be observed as a function of temperature and exchanges rate within the range of ca. 10^4 s^{-1} to 10^8 s^{-1}, which constitute the slow and fast exchange limits, respectively.

Lineshape simulations in the intermediate exchange regime require the input of several parameters [48]. The QCC constant can be experimentally determined from spectra measured at very low temperatures, when the sample is static. The simulation requires a suitable exchange model for the system that accounts for the trajectories of motion. Examples include 180° rotations for phenylene groups, 120° rotations for methyl groups, etc. Other parameters include the populations of sites with different energies, any asymmetry in the electric field gradient tensor (η), and, most importantly, the site exchange

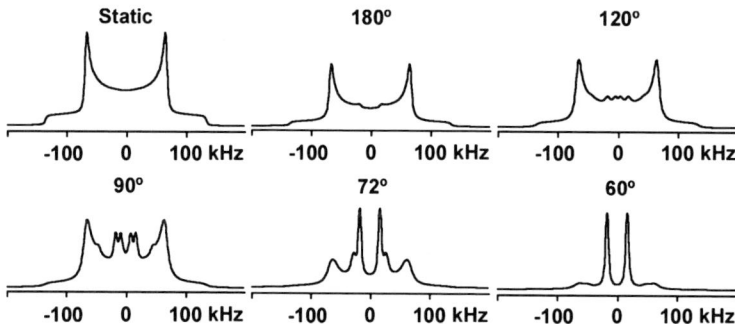

Fig. 8 Simulated ^{2}H NMR line shapes for a static rotator, and for rotators undergoing jumps with angular displacements of 180°, 120°, 90°, 72°, and 60° with an exchange rate of 10^5 s^{-1}

rates. These measurements are taken at several different temperatures and the energetics of the system determined by Arrhenius analysis of the exchange rate vs. temperature.

3.4
Spin-Lattice Relaxation

Spin lattice relaxation measurements can also be used to probe molecular and lattice dynamics in crystalline solids by following changes in the relaxation rate (T_1^{-1}) as a function of temperature (T). Spin lattice relaxation is the time that it takes for a given nucleus to attain equilibrium at a given temperature under some external magnetic field [49, 50]. Interestingly, spontaneous transitions between nuclear states are so unlikely that they would take millions of years. In fact, nuclear transitions required to reach thermal equilibrium are stimulated by random magnetic fields generated by nearby nuclear spins, rotating dipoles, and moving charges. Since it is well known that spin-lattice relaxation in solid samples may be dominated by a particular dynamic process, it is a very promising method to determine the dynamics of molecular gyroscopes where portions of the lattice are fairly static while others experience rapid dynamics. In ideal cases, a minimum in the relaxation rate is observed when the rate of the dominant dynamic process matches the Larmor frequency of the NMR nucleus being observed (Fig. 9). For example, ^{1}H T_1 relaxation experiments carried out at 300 MHz will show a minimum at a temperature where the dynamics of the sample have a rate of 300 MHz. If observations are made for a ^{13}C signal, the minimum T_1 value will occur at a temperature where the rate matches the corresponding Larmor frequency of 75 MHz. If a single dynamic process with a correlation time τ_c is indeed responsible for the T_1 relaxation within a given temperature regime, one can take advantage of the Kubo–Tomita relation Eq. 3 [51] to determine the rate constant of this process ($k = 1/\tau_c$).

$$T_1^{-1} = C[\tau_c(1 + (\omega_0^2 \tau_c^2))^{-1} + 4\tau_c(1 + 4(\omega_0^2 \tau_c^2))^{-1}] \tag{3}$$

$$\tau_c = \tau_0 \exp(-E_a/kT) \tag{4}$$

The constant C in Eq. 3 is proportional to the number of protons, ω_0 is the nuclear precession frequency, and τ_c the correlation time for the dynamic process causing relaxation. By fitting the log of the relaxation rates vs. the inverse temperature to Eqs. 3 and 4, one obtains the value of τ_0, E_a, and C as adjustable variables. Relaxation measurements can also be used to test for the existence of phase transitions that may change the dynamics of the sample and its relaxation rate. If the sample goes through a phase transition in the temperature range under study, a discontinuity is observed at the phase transition temperature in the T_1 vs. temperature curve.

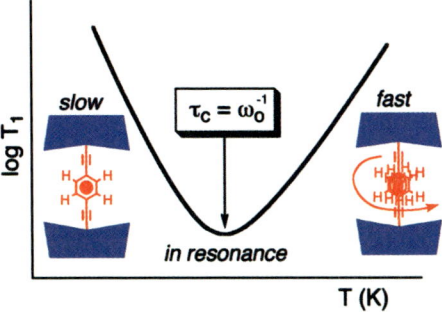

Fig. 9 Variation in spin-lattice time constants as a function of temperature for a solid sample with a dominant dynamic process with a correlation time τ_c in a spectrometer that operates at a frequency ω_0 that corresponds to the nucleus being observed

4
Molecular Gyroscopes with Triptycyl Stators

Triptycenes are rigid structures with three benzene rings fused with a bicyclo[2.2.2]octatriene. As illustrated in Fig. 10, two oppositely positioned triptycyl moieties provide a potentially suitable scaffold with the proper directionality to connect an axle linked to a relatively small rotator. The

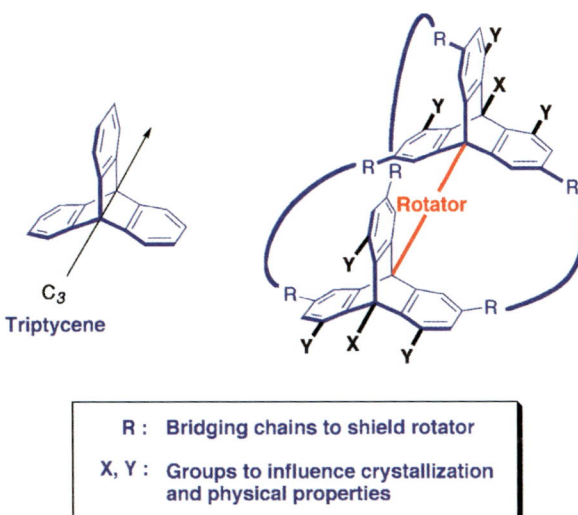

Fig. 10 (*left*) Structure of triptycene illustrating the three-fold symmetry axis along the direction where the rotators can be attached, and, (*right*) structure of a molecular gyroscope with a bis-triptycyl stator illustrating the position of substituents used to close the frame and to change its physical properties

triptycene frame provides opportunities for functionalization with bridging chains (-R-R-) to encapsulate the rotator, and to attach substituents that may help manipulate the crystallization and physicochemical properties of the bulk materials. We have formulated the synthesis of molecular gyroscopes with symmetric triptycyl stators and aromatic rotators in terms of relatively simple and highly convergent strategies involving sequential Diels–Alder reactions and Pd(0)-catalyzed coupling of terminal alkynes with aryl halides [52]. These are illustrated in Fig. 11.

The intrinsic (gas phase) rotational dynamics of molecular gyroscopes with triptycyl stators should be determined by the energetics of rotation about the sp-sp^3 and sp-sp^2 bonds or the triptycyl-alkyne and alkyne-arene single bonds, respectively. Unlike the extremely well documented rotation about single bonds involving sp^3 and sp^2 hybridized carbon atoms [53], much less is known about the energetics of rotation about single bonds involving sp-hybridized carbons. However, as expected from the cylindrical symmetry of alkyne π-systems, experimental and computational studies suggest that rotation of groups attached to acetylenes should be essentially barrierless [54–56].

The gyroscopic dynamics of the central rotator will also depend on the number of conformational states that can be explored with sufficient thermal activation. Ideal molecular gyroscopes would possess perfectly flat rotational potentials with rotators that can adopt an infinite number of orientations

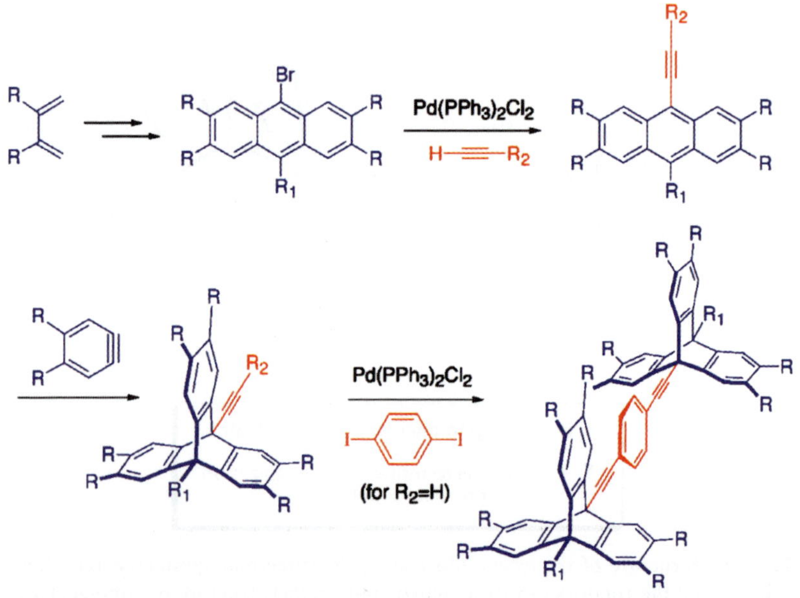

Fig. 11 General synthesis of molecular gyroscopes with triptycyl stators and aromatic rotators from simple butadienes and dihaloaromatics

about their rotational axes. However, real rotators will have orientation-dependent energies with energy minima and maxima arising from through bond and through-space effects, including electrostatic interactions and weak dispersive forces. In practice, it should be the number and relative energies of these conformational states, which should determine the internal dynamics of the isolated molecular rotator.

The conformational space of aryl-substituted molecular gyroscopes with triptycyl frames can be defined in terms of the dihedral angles formed by the planes of the two triptycyl groups with respect to each other, and with respect to the plane of the central arylene [52] (Fig. 12). In analogy to ethane, we could illustrate the relative position of the two triptycyl groups by projections viewed along an axis passing through the triptycene bridgehead carbons, as shown in Fig. 12. In principle, the triptycyl groups may be eclipsed E(Θ), staggered S(Θ), or adopt anyone of an infinite number of structures in between. In the meantime, the central rotator may adopt a dihedral angle Θs, given by the planes of the arylene rotator and the plane of a benzene ring in a reference triptycene [52].

The number of conformational states for molecular gyroscopes with triptycene stators and flat aromatic rotators can be interpolated from the number of symmetry-related structures that can be identified with respect to the molecular axis corresponding to the axis of internal rotation. The highest rotational symmetries available to the linearly-linked triptycyl stators are a C_3-axis and an S_6-axis for E(Θ) and S(Θ) conformations, respectively. In contrast, the flat aromatic rotators can only have a 2-fold axis. When the aromatic rotator eclipses one of the aromatic rings of a triptycyl group, three degenerate structures are generated when the two triptycenes are eclipsed [E(0), E(120) and E(240)] and six degenerate structures when they are staggered [S(0), S(60), S(120), S(180), S(240), and S(300)]. Similarly, when the plane of the aromatic rotator is orthogonal to the plane of the reference benzene rings of the two triptycenes, there are three degenerate structures for the eclipsed

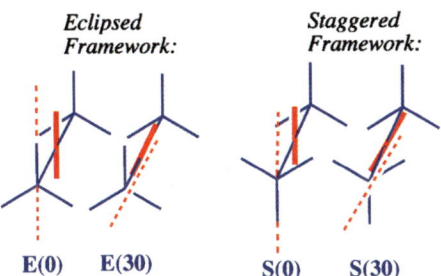

Fig. 12 The two potential orientations of the triptycyl groups with respect to each other: eclipsed (*left*) and staggered (*right*). For each of these orientations, two positions are illustrated for the rotator [52]

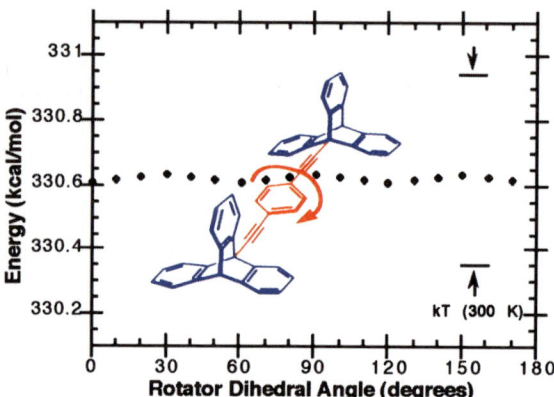

Fig. 13 Gas-phase changes in energy (AM1 method) in 1,4-bis-[2-(9-triptycyl)-ethynyl]benzene as a function of phenylene rotation in a stator of eclipsed triptycyls. The minimum corresponds to E(0) and the maximum to E(30)

frame [E(60), E(180) and E(300)] and six for the staggered one [S(30), S(90), S(150), S(210), S(270), and S(330)]. Thus, even without knowing whether the eclipsed and/or bisected conformations of the rotator are energy minima or maxima, one may realize that eclipsed stator conformations will have half as many conformers as the structures where the two components of the triptycyl stators are staggered. Some insight was obtained from simple calculations with the AM1 method for 1,4-bis-[2-(9-triptycyl)-ethynyl]benzene, one of the simplest model structures for this class of compounds [52]. Energy minimizations that started with arbitrary dihedral angles between triptycyl and phenylene groups ended locally with energy differences that were more than 0.05 kcal/mol. Varying the relative orientation of the two triptycyl groups made no difference on the size of the calculated energies. Conformations E(0) and S(0) are predicted to be within 0.004 kcal/mol. The energy profile obtained by rotation of the phenylene group with a framework that has the two triptycenes eclipsed, E(Θ), is shown in Fig. 13 along with marks that indicate the value of kT for $T = 300$ K. While the accuracy and resolution of these computations is clearly questionable, there is little doubt that rotation of the triptycyl and phenylene groups in the ground state is essentially barrierless and that gyroscope motion would be limited by the moment of inertia of the rotator for molecules in the gas phase [52].

4.1
A Molecular Gyroscope with a bis-Triptycyl Stator and a Phenylene Rotator

Samples of 1,4-bis-[2-(9-triptycyl)-ethynyl]benzene **1** provided a good starting point to test the synthesis and to set a benchmark to understand the physical properties of this class of compounds. One of the most relevant characteristics of **1** was its very poor solubility in aliphatic solvents (hexanes, dichloromethane, chloroform, and acetone [52]), and its limited solubility in aromatic solvents like benzene, toluene, and *meta*-xylene. Single crystal X-ray diffraction structural determination of molecular gyroscope **1** was carried out after obtaining very elusive X-ray-quality single crystals from *meta*-xylene. The structure of **1** was solved as a crystalline solvate in the triclinic space group P1-bar with one molecule of **1** and one of *meta*-xylene per unit cell. It was found that the two triptycyl groups in the structure of **1** adopt a nearly eclipsed conformation with the central phenylene group being also coplanar in a E(0) conformation (Fig. 14). The acetylene groups are almost collinear with a deviation of only $1.1°$.

The packing structure of compound **1** is characterized by having all molecules with their long molecular axes parallel, as expected for a crystal

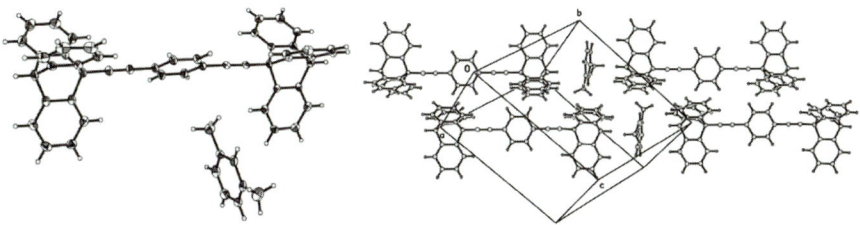

Fig. 14 (*left*) ORTEP diagram of the asymmetric unit of rotor **1**, which includes one molecule of *meta*-xylene. (*right*) Packing diagram of compound **1** illustrating a unit cell with all molecules aligned in the same direction

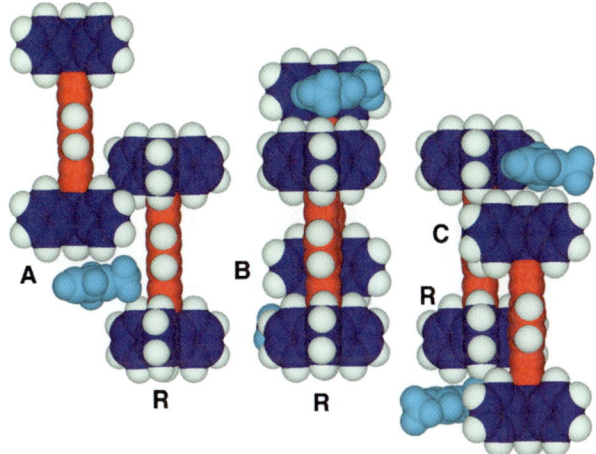

Fig. 15 Close packing interactions between close neighboring molecules labeled **A**, **B**, and **C** and the phenylene rotator of a reference molecule labeled **R**. The location of a solvent molecule is also illustrated [52]

formed with rigid rods (Fig. 15). As expected from its open structure, there are several contacts between the phenylene rotator of **1** and its surroundings. These are illustrated in three views of a reference molecule (R), its three closest neighbors (A–C), and the *meta*-xylene molecule depicted in Fig. 15. A very tight interdigitation between molecules of **1** can be appreciated in the close contacts shown in the figure between molecules **RB** and **RC**. Thermal analysis of the clathrate revealed the loss of *meta*-xylene between 120 and 160 °C. Notably, crystals of **1** do not melt up to ca. 400 °C, at which point the sample begins to decompose. Given the tight crystal packing near the central rotator, it was expected that crystals of **1** should be static. This was qualitatively confirmed by solid-state ^{13}C CPMAS-DD NMR, where all protonated carbon signals disappeared upon dipolar dephasing, except the rapidly rotating methyl groups of *meta*-xylene.

4.2
Towards Structural Refinements in Triptycyl Stators

In order to create true amphidynamic crystals using a triptycyl stator and phenylene rotator, one must prevent interdigitation of the adjacent triptycenes into the pocket of rotation. One possible solution would be the addition of bulky substituents (-R) to the peripheral carbons of the two triptycenes, as shown in Fig. 16 [57]. Additional substitutions at the two bridgehead positions may be used to improve solubility and other physical properties.

Simple methyl groups were selected as a starting point to test the effect of peripheral substituents on the crystallization of the triptycene stators. However, several challenges arose during the preparation of these compounds. Among these, it was found that the methyl-substituted anthracenes and anthranylic acids required in their synthesis, while relatively easy to prepare,

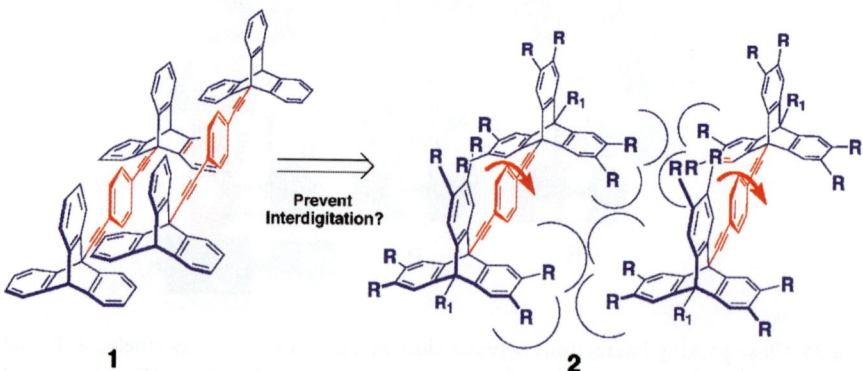

Fig. 16 Proposed substitution pattern of the triptycyl frame to prevent interdigitation [57]

were exceedingly insoluble. Furthermore, the Diels–Alder reaction between methyl-substituted benzynes and anthracenes proceeds with low regioselectivity, forming an undesirable napthobarrelene adduct by reaction at one of the lateral rings of the substituted anthracene. To improve sample solubility and Diels–Alder selectivity, several triptycenes were studied with substituents at the bridgehead position, and compounds **2**, **3**, and **4**, with methyl, phenyl, and propyl groups, respectively, were carried through to the final molecular gyroscopes (Fig. 17) [57].

The precursors required for the synthesis of compound **2** were barely tractable and the overall yields were low. The final molecular gyroscope itself was exceedingly insoluble and essentially intractable. All attempts to obtain single crystals resulted in white microcrystalline powders. The inclusion of a phenyl group at the bridgehead position was only analyzed in a structure with eight rather than twelve peripheral methyl groups in compound **3**. The main limitation in the synthesis of **3** was the low selectivity of the Diels–Alder reaction to form the required ethynyl triptycene. Since the phenyl group in compound **3** improved its solubility only marginally, the synthesis of the dodecamethyl derivative was not pursued further. Finally, the small size and added flexibility of the propyl group in compound **4** was an excellent solution for a convenient synthesis and dramatic improvement in solubility and crystallinity. Single crystal X-ray diffraction studies of the phenyl (**3**) and propyl (**4**) derivatives showed their tendency to include solvents of crystallization.

The crystal structures of molecular gyroscopes **3** and **4** were solved in the space group P1-bar. The phenyl-substituted **3** crystallized in two slightly different molecular structures with two molecules of *meta*-xylene per unit cell. Both structures of **3** have triptycyl groups in a staggered conformation with their phenylene rotator having a rotational angle that can be labeled as S(23.2°) and S(56.7°). Similarly, the propyl substituted compound **4** crystallizes with two molecular gyroscopes related by a center of symmetry and four molecules of bromobenzene per unit cell. When viewed along the direction

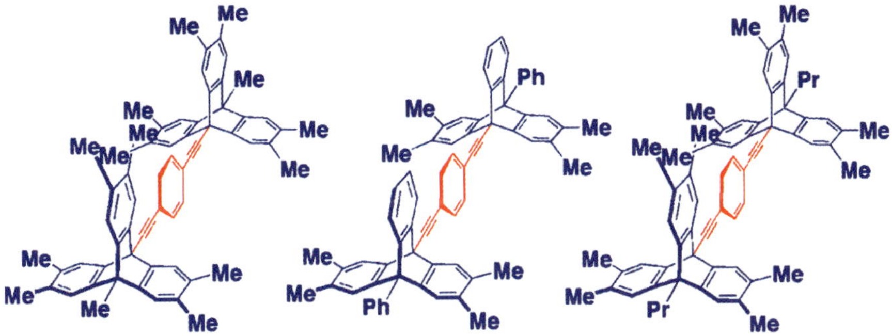

Fig. 17 Molecular gyroscopes with methyl-substituted triptycyl stators

of the phenylene rotational axis, the two triptycyl groups adopt a staggered conformation with the rotator almost coplanar with two triptycyl planes [S(4.4°)]. Both compounds **3** and **4** have structures that deviate slightly from linearity as a result of a small bending of their triple bonds.

The packing structures of **3** and **4** both have all the molecules in the crystal aligned in the same direction (Fig. 18). The close interdigitation between nearest-neighbors previously observed in a crystal of the parent compound **1** was modified into a packing structure that separates the phenylene groups of adjacent molecules by $\sim$ 6.7–7.5 Å. Not surprisingly, the separation of adjacent molecules creates large cavities that are filled by the *meta*-xylene molecules. The phenylene rotators are "sandwiched" between two molecules of *meta*-xylene in a parallel co-facial interaction in one of the structures, and in a non-parallel, C – H–π, edge-on relation with the two molecules of *meta*-xylene in the other [57].

The packing structure of the propyl derivative **4** is essentially the same as the one proposed in our original design (Fig. 16). The methyl groups on the periphery of the two triptycenes prevent interdigitation between adjacent molecular rotors, thus creating relatively large cavities between them. As expected for a relatively rigid rod, all the molecules in a crystal are oriented in the same direction. Translation of molecular gyroscopes in two directions perpendicular to their molecular long axes results in the formation of layers

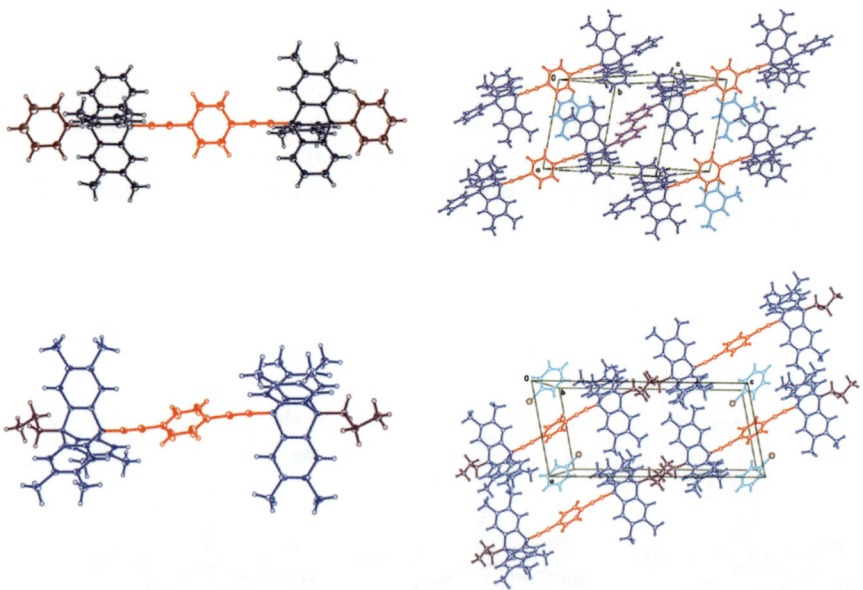

Fig. 18 (*left*) ORTEP diagram of molecular gyroscopes **3** and **4** (*top* and *bottom*, respectively) with thermal ellipsoids at the 50% probability level and the solvent removed for clarity. (*right*) Partial packing diagrams

that segregate from each other by the propyl groups at the two bridgehead positions. While the local environment around the central phenylene of each molecular rotor is shared by six bromobenzene molecules, the thermal ellipsoids for the carbons of the rotor are clearly elongated in a direction perpendicular to the plane of the phenylene ring, even at 100 K, suggesting the wide amplitude phenylene librations expected when there is rapid gyroscopic motion. Preliminary solid-state NMR studies indicate exceedingly fast rotation rates of ca. 10^7 s^{-1}, even at 273 K (– 100 °C). These results indicate a dramatic improvement when compared to the unsubstituted molecular rotor 1, as well as the realization of a de novo designed amphidynamic crystalline material [58].

5
Molecular Gyroscopes with Triphenylmethyl Stators

Removal of the external bridgehead carbons from the triptycyl stators and replacement by hydrogen atoms releases six rotational degrees of freedom and transforms them into the analogous triphenylmethyl (trityl) stators (Fig. 19). It is expected that six relatively soft C-Ph conformational degrees of freedom will render the trityl stators significantly more soluble, and with some shape flexibility that may help optimize its packing interactions. In addition, trityl groups are known to adopt propeller conformations which endows them with helical chirality, thus raising the possibility for the stators to crystallize in chiral or meso forms, simply depending on the sense of helicity of the two propeller groups. The robust crystallization properties of many compounds with trityl groups have led to the suggestion that they can be used as a supramolecular synthon for crystal design [59, 60]. It has been shown that trityl derivatives tend to adopt complementary edge-

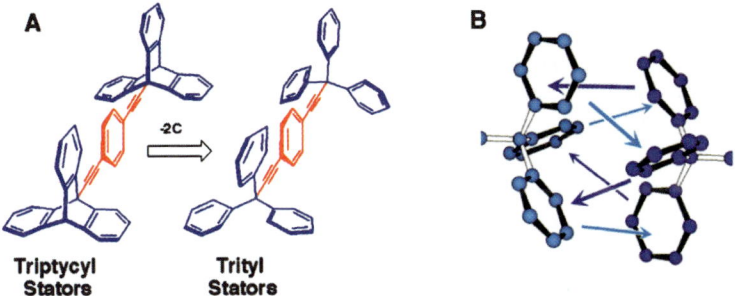

Fig. 19 A Transformation of a triptycyl stator into a trityl stator by removal of the two distal bridgehead carbons. **B** Illustration of a six-fold trityl embrace consisting of six complementary edge-to-face interactions between adjacent trityl groups

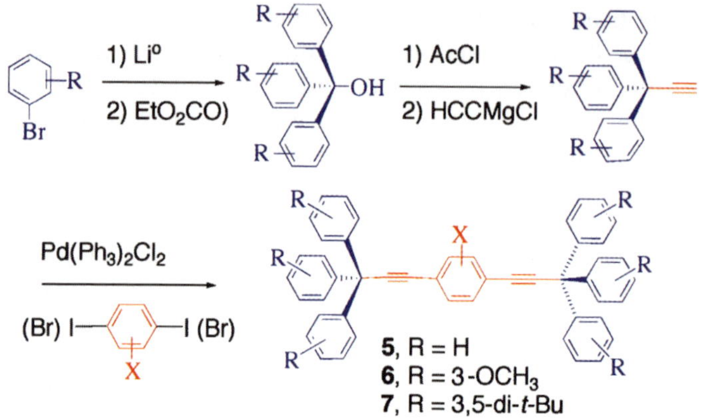

Fig. 20 Three-pot general synthesis of molecular gyroscopes with trityl stators

to-face interactions with enantiomerically related propellers around a local S_6 symmetry axis, in a structure that has been termed the six-fold phenyl embrace (Fig. 19b).

A highly versatile and efficient synthetic strategy for the preparation of symmetric trityl stators with aromatic rotators may be formulated in only three "pots" from bromobenzenes as illustrated in Fig. 20 [61]. Following the preparation of symmetric trityl alcohols by addition of aryl lithium reagents with diethyl carbonates, the triple bonds can be conveniently added by a substitution reaction of an intermediate trityl chloride with ethynyl magnesium chloride. The aromatic rotator can be added as a substituted dihalo phenylene in a double Pd(0)-coupling reaction [62].

5.1
A Molecular Gyroscope with a bis-Triphenylmethyl Stator and a Phenylene Rotator

The parent trityl structure **5** was prepared in reasonably good yields in large quantities from commercial trityl chloride, ethynyl magnesium bromide, and diiodobenzene [61]. Diffraction quality single crystals obtained from benzene and from methylene chloride were shown to be pseudopolymorphic, a term that has been adopted to describe different crystal forms of a given compound differing in stoichiometry, in this case due to the presence of solvent of crystallization [63]. Crystals grown from benzene formed a cage with the solvent, or clathrate, with a structure that may be described in terms of four molecules of **5** acting as pillars that entrap a parallel displaced benzene in the center [61, 64]. A clear similarity may be appreciated between compound **5** and the so-called wheel-and-axle complexes reported by Toda and co-workers, which are also known to entrap small molecules in their crystal lattice [65, 66]. Similar to the triptycyl structures, both pseudopolymorphs of **5** have all the molecules packed with their long molecular axis aligned in the same direction (Fig. 21), forming infinite chains of molecules interacting by a six-fold trityl embrace [64]. Thermogravimetric analysis (TGA) and differential scanning calorimetry (DSC) were used to determine the thermal stability of the two structures. DSC analysis of the benzene clathrate showed an endothermic transition at ca. 100 °C followed by an exothermic peak at 105 °C and a sharp melting peak at 316 °C. TGA and microscopic analysis confirmed that the transition at 100–105 °C corresponds to desolvation and subsequent recrystallization in a solid-to-solid phase transition. The melting at 316 °C was shown to be identical for both the thermally desolvated clathrate and the solvent-free crystals from methylene chloride. Their phase identity could be confirmed spectroscopically by ^{13}C CPMAS both with samples prepared independently, and in situ, with experiments carried out as a function of temperature where desolvation and the corresponding phase transition can be clearly observed (Fig. 22).

While both crystal structures of **5** are close packed and reveal no open cavities, a dynamic analysis by ^{13}C CPMAS coalescence and wide-line ^{2}H NMR analysis revealed a relatively efficient gyroscopic motion [61, 64]. With favorable resolution and chemical shift dispersion of the phenylene rotor hydrogens, the benzene clathrate was amenable to coalescence analysis.

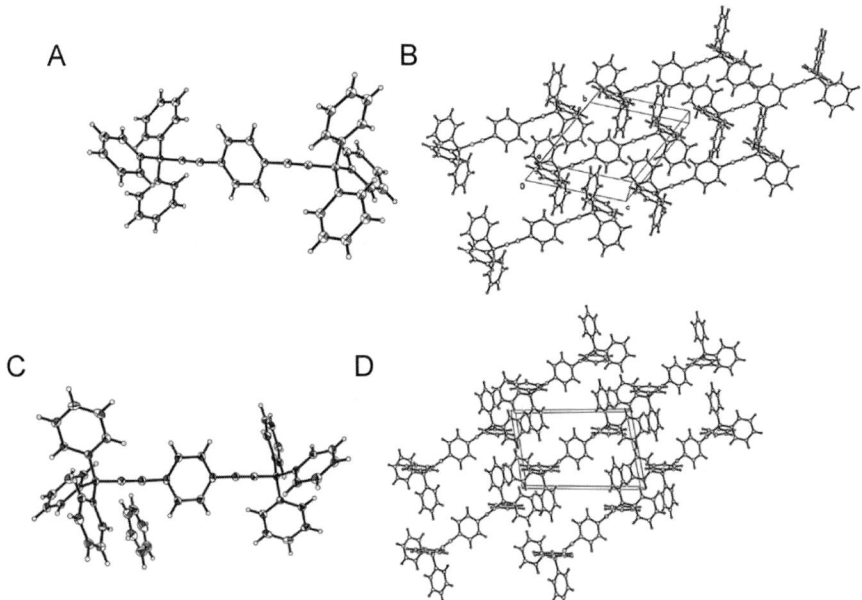

Fig. 21 **A** ORTEP diagram of thermal ellipsoids drawn at the 30% probability and **B** packing diagrams of the trityl rotor **5** crystallized from CH_2Cl_2. **C** and **D** Equivalent diagrams from the crystal structures obtained from benzene (benzene molecules were removed for clarity)

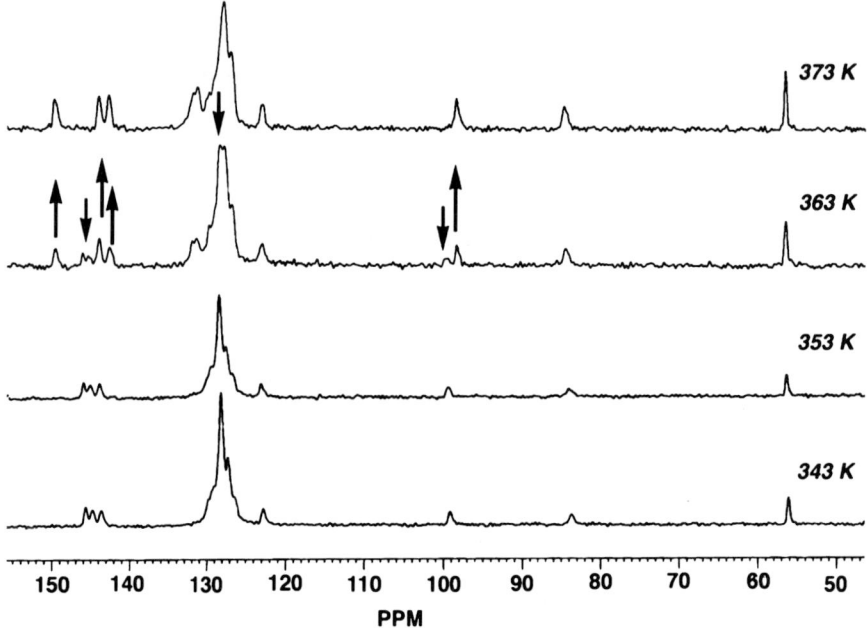

Fig. 22 ^{13}C CPMAS NMR spectra of a freshly crystallized sample of benzene clathrate of **5** before (*bottom*) and after heating at the temperature indicated. The spectrum obtained after heating the sample to 373 °C is identical to that obtained from crystals grown in methylene chloride [64]

In a variable temperature ^{13}C CPMAS study between 220 K and 297 K, signals corresponding to the protonated carbons of the rotator at 131.6 and 130.8 ppm were shown to coalesce at ca. 255 K, by a dynamic process that involves rotation about the alkyne axle (Fig. 23, signals marked with asterisks). Using a single point approximation from the coalescence temperature and the difference in chemical shift between the two exchanging signals [44], one may calculate a rotational barrier of 12.8 kcal/mol and a time constant of 7.7 ms (rate of 130 s^{-1}).

In contrast to the benzene clathrate, the phenylene signals of the solvent free structure had a severe spectral overlap with those of the trityl frame. To remove this spectral interference we devised a strategy for selective ^{1}H–^{13}C cross polarization of the rotator by preparing samples of **5-d$_{30}$**; an isotopolog of **5** with a protonated rotator and a per-deuterated trityl frame [67]. This strategy takes advantage of the distance-dependence of the cross polarization rate and efficiency. While ^{1}H magnetization transfer occurs very rapidly to covalently linked carbons, magnetization transfer occurs much slower to non-bonded carbon atoms that are farther away. Figure 24 illustrates the aromatic region of spectra recorded with **5-d$_{30}$** using cross polarization times of 8 ms (top) and 50 µs (bottom). While all carbon signals are visible in the

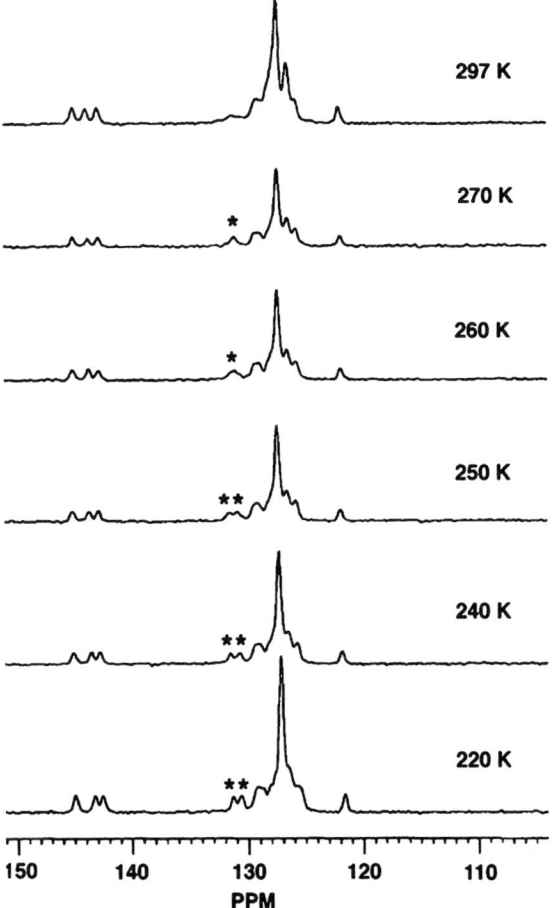

Fig. 23 Variable temperature ^{13}C CPMAS NMR spectra of the benzene clathrate of **5**. The signals corresponding to the phenylene signals involved in the coalescence process are indicated with asterisks [64]

top spectrum as polarization transfer has been allowed to approach equilibrium, only the protonated signals of the phenylene rotator are visible in the bottom spectrum when polarization transfer has been allowed for a very limited time. Line-shape simulations [67] of variable temperature spectra from samples of solvent free crystals of **5**-d_{30} acquired with short contact times provided exchange rates from nearly static at 214 K to 1.2 KHz at 308 K (Fig. 25). An Arrhenius plot constructed with the rate and temperature data provided an activation barrier of 11.3 kcal/mol and a pre-exponential factor of 2.9×10^{11} s^{-1}.

Variations in rotational dynamics by ^{13}C CPMAS coalescence analysis observed between the solvent free and the benzene clathrate are rather small

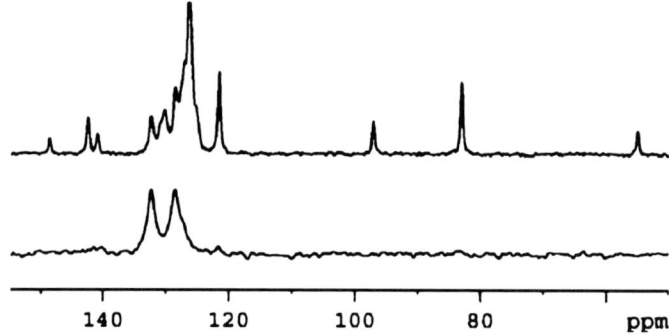

Fig. 24 ^{13}C CPMAS NMR spectrum of 5-d$_{30}$ with a contact time of 8 ms (*top*) and 50 μs (*bottom*) at 214 K. The longer contact times allow for transfer of polarization to all carbon atoms in the structure. Very short contact times limit magnetization transfer to the phenylene carbons that are directly bound to the only ^{1}H in the structure [67]

and probably within the experimental error of the methods. Recognizing the high thermal stability of the solvent free structure it was of interest to investigate its internal rotational motion within the much faster dynamic range (ca. 10^4–10^8 s^{-1}) provided by the ^{2}H NMR method. To that effect, samples of 5-d$_4$ had to be prepared with a deuterated rotator and a fully protonated stator. Measurements carried out with the desolvated sample between 297 and 385 K showed spectral changes consistent with a dynamic process described by rotation of the phenylene group (Fig. 26). The spectra were simulated with

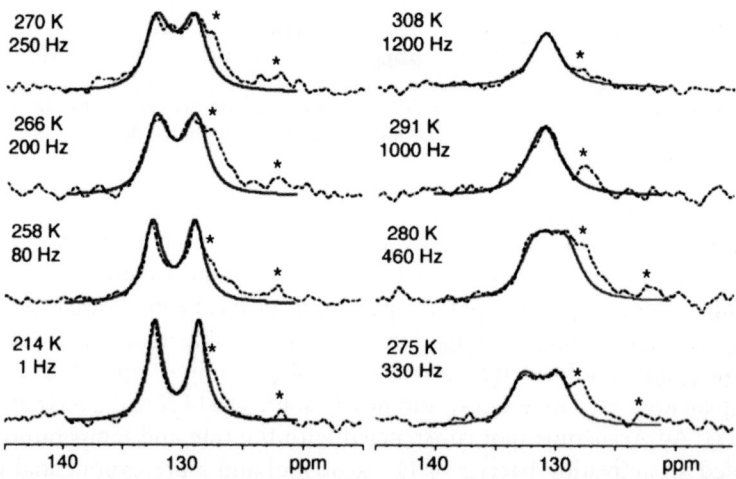

Fig. 25 Variable temperature ^{13}C CPMAS NMR of 5-d$_{30}$ with a contact time of 50 μs (*dotted line*), and simulations from g-NMR [45] (*filled line*). The stars indicate residual signals from the non-protonated phenylene carbon and one of the trityl signals [67]

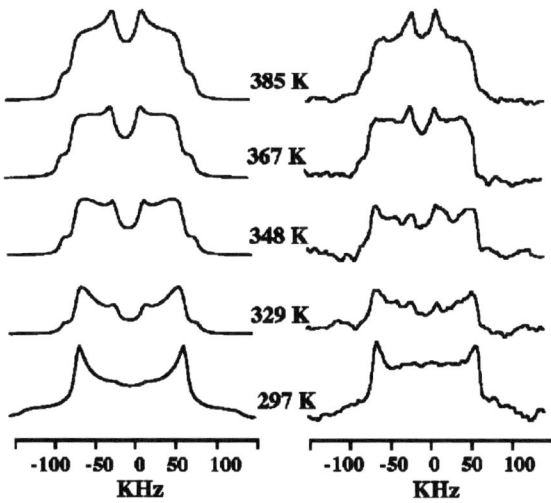

Fig. 26 Experimental (*right*) and simulated (*left*) solid state ^{2}H NMR of desolvated samples of 5-d$_4$. The rotation rate constants used for fitting, from *bottom* to *top* are ($\times 10^6$ s^{-1}): 0.015, 0.4, 1.3, 2.2, and 3.8 [61, 64]

a model that assumes a quadrupole coupling constant QCC = 180 KHz and an anisotropy parameter $\eta = 0$. The model also assumes a two-fold flipping motion (180° rotations) with exchange rates that varied from 1.5×10^4 to 3.8×10^6 s^{-1} within the temperature limits analyzed (Fig. 26) [64]. An Arrhenius plot gave a barrier to rotation of 14.6 kcal/mol, which is 2–3 kcal/mol higher than those determined by the ^{13}C CPMAS coalescence method. These differences may be the result of activation barriers that are different at different temperatures, of the different isotopologues used in all three experiments (1, 1-d$_4$, and 1-d$_{30}$), or the accumulated experimental errors of these techniques [67].

5.2
Structural Diversity Through Polymorphism in Molecular Gyroscopes with Substituted Trityl Stators

Variations in the structure of molecular gyroscopes are expected to cause changes in their crystal packing and rotary dynamics. In order to establish clear structure-activity relations, the X-ray structures and dynamic data from a large number of compounds in a homologous series would be desired. While this information may be obtained by taking advantage of synthetic strategies to introduce a large variety of substituents, molecular gyroscopes with substituted trityl frames are conformationally very rich and may be ideally suited to form a large number of crystal forms from any given compound [68]. In fact, it is known that crystallization relies on a subtle balance of inter- and intra-molecular forces and that lattice energies of polymorphic crystals typically differ by ∼ 1–2 kcal/mol [63, 69]. With that in mind, we have recently suggested that substituted trityl frames with a large number of conformers within a small energy range should be prone to form many polymorphs and pseudo-polymorphs [70, 71]. To illustrate the remarkably rich conformational landscape of molecular gyroscopes with trityl stators we recently analyzed the conformational and crystallization tendencies of 1,4-bis[tri-(*meta*-methoxyphenyl)propynyl]benzene **6** (Fig. 27).

While apparently very simple, compound **6** has a very large number of geometric isomers (Fig. 27). To appreciate this number one must start by considering the conformational preferences and static stereochemistry of substituted trityl derivatives [72]. In addition to the chiral propeller conformations resulting from the uniform tilt of the planes of the three rings with respect to the Ar$_3$C – X bond in each trityl group [73], aromatic substituents lacking a local two-fold symmetry axis coincident with the direction of their Ar – C bond possess a stereogenic element known as planar chirality. In the case of the *meta*-substituted benzenes of compound **6** this can be easily recognized in terms of the orientation of the MeO-substituent (either syn- or anti-) with respect to the rotator at the center of the molecule. As illustrated in Fig. 27, two stereoisomers are obtained from a given trityl group configuration when the orientation of the methoxy group is considered. Moreover, gas phase and computational analysis of simple *meta*-methylanisol and other *meta*-methoxy aromatics indicate the existence of co-planar syn- and anti-conformers with an energy difference of only ca. 60 cm^{-1} (0.17 kcal/mol) [74]. Thus, the geometric isomers generated by the orientation of each methyloxy group with respect to each aromatic *ipso*-carbon become an additional stereogenic element that increases the number of minima in the conformational landscape of

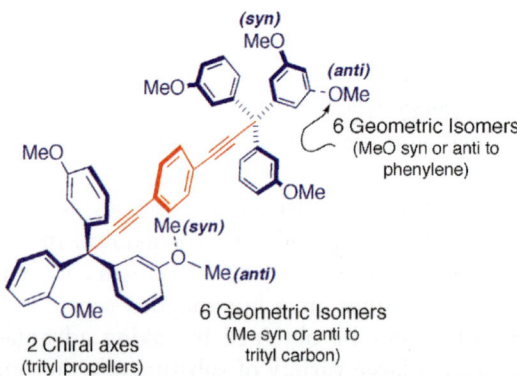

Fig. 27 Line formula of 1,4-bis[tri(*meta*-methoxyphenyl)propynyl]benzene **6** illustrating its chiral (trityl propeller) axes, and the geometric isomerism (syn-anti) of the *meta*-methoxy and methyl groups

6. In addition, the relatively flat torsional potentials of the central phenylene and six aryl groups may give rise to non-identical structures differing only in the magnitude of one or more dihedral angles between the various aromatic planes [75]. A systematic Monte Carlo multiple minimum conformational analysis of 15 000 structures revealed ca. 1608 different structures within 2 kcal/mol, and as many as 4641 within energy windows of 5 kcal/mol [68].

In agreement with our expectations, a small number of crystallization experiments resulted in a relatively large number of crystal forms, which were identified by DSC, single crystal X-ray diffraction, solid-state ^{13}C CPMAS NMR and microscopic observations. As indicated in Fig. 28, phases labeled A, B, and G were fully characterized by single crystal X-ray diffraction analysis. Crystal phases A, F, and G were shown to be benzene, acetone, and dichloromethane clathrates, respectively (Fig. 30). Crystal forms B, C, D, and E are solvent free and were formed from the other pseudopolymorphs by desolvation and recrystallization, or from other polymorphs by recrystallization, as indicated by the arrows in Fig. 28. The transformation sequence A → B → C → D was documented in-situ by ^{13}C CPMAS NMR. Spectral changes, in the form of a growing number of signals, suggested lower molecular symmetry or a larger number of conformers in crystals formed at higher temperature (Fig. 29).

X-ray analysis showed that the structures of phases A, B, and G are centrosymmetric with coincident crystallographic and molecular inversion centers. The structure of the solvent-free crystal (B) was solved in the monoclinic

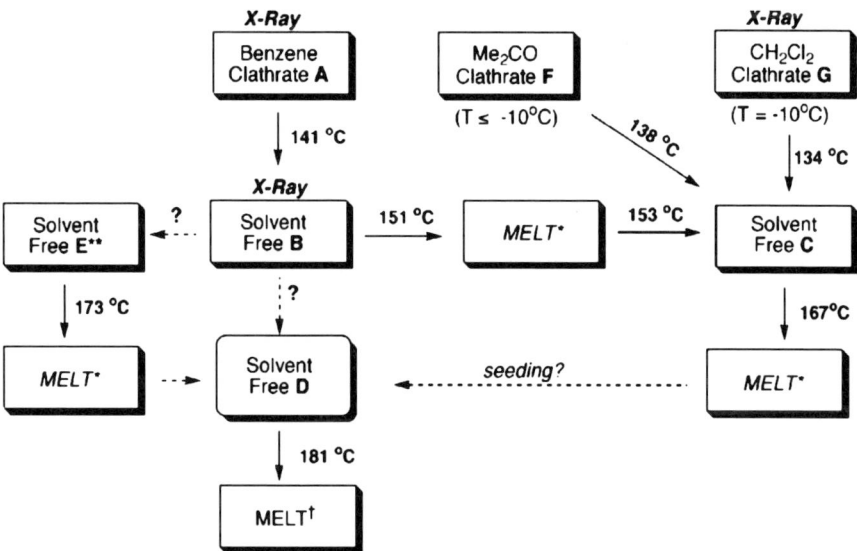

Fig. 28 Block diagram illustrating the known polymorphic and pseudopolymorphic phases of the hexamethoxy-molecular gyroscope **6** [68]

space group $P2_1/c$, while those of clathrates A and G were solved in the triclinic space group P1-bar. As expected, molecular gyroscope **6** occurs in three different conformers in the three crystal structures. In form A, one methoxy group adopts an endo-conformation, pointing towards the central phenylene rotator, while the other two adopt the exo-forms. The structure of **6** in form B has all methoxy substituents endo-, but two of them have methyl groups oriented towards the *para*-phenyl carbon and the third one towards the *ortho*-position. The structure of **6** in form G has all of the methoxy substituents in an endo-conformation, two methyl groups towards the *ortho*-carbon, and the third one toward the *para*-position.

We believe that the structural richness of molecular gyroscopes with substituted trityl frames will translate into many opportunities to investigate the relation between structure and gyroscopic dynamics, and will offer opportunities for the design of interesting materials. The introduction of three different aryl groups will render the trityl carbon intrinsically chiral, adding one more stereogenic element that may have a deep influence on the rest. Molecular gyroscopes with chiral stators suggest many interesting properties and their preparation and study are currently under way in our laboratories.

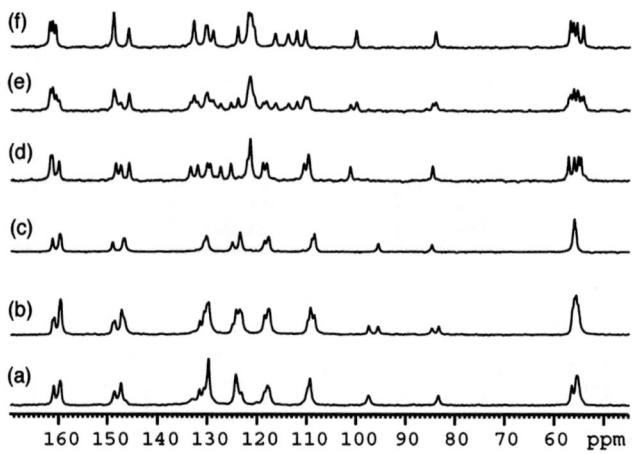

Fig. 29 Solid state CPMAS 13C NMR spectra of **6** taken at 298 K after having exposed the sample for 30 min periods to reduced pressure (10^{-3} torr), or ca. 10 degree temperature increments, until the change was complete: **a** Samples freshly crystallized from benzene remained unchanged for no less than a week at 298 K (phase A). **b** Some desolvation occurs after exposure to 10^{-3} torr for 6 h (phases A and B). **c** Pure phase B was obtained after heating phase A to 135 °C for 60 min. This spectrum was identical to that obtained with crystals grown from CH_2Cl_2. **d** Phase C is obtained after heating to 153 °C for a total of 3 h. **e** Phases C and D can be detected after heating to 160 °C for a total of 3 h. **f** Pure phase D after heating to 160 °C for a total of 12 h. No changes corresponding to phase E were observed and melting occurred at 180 °C. The spectrum obtained after melting consisted of very broad peaks as expected for an amorphous glass [68]

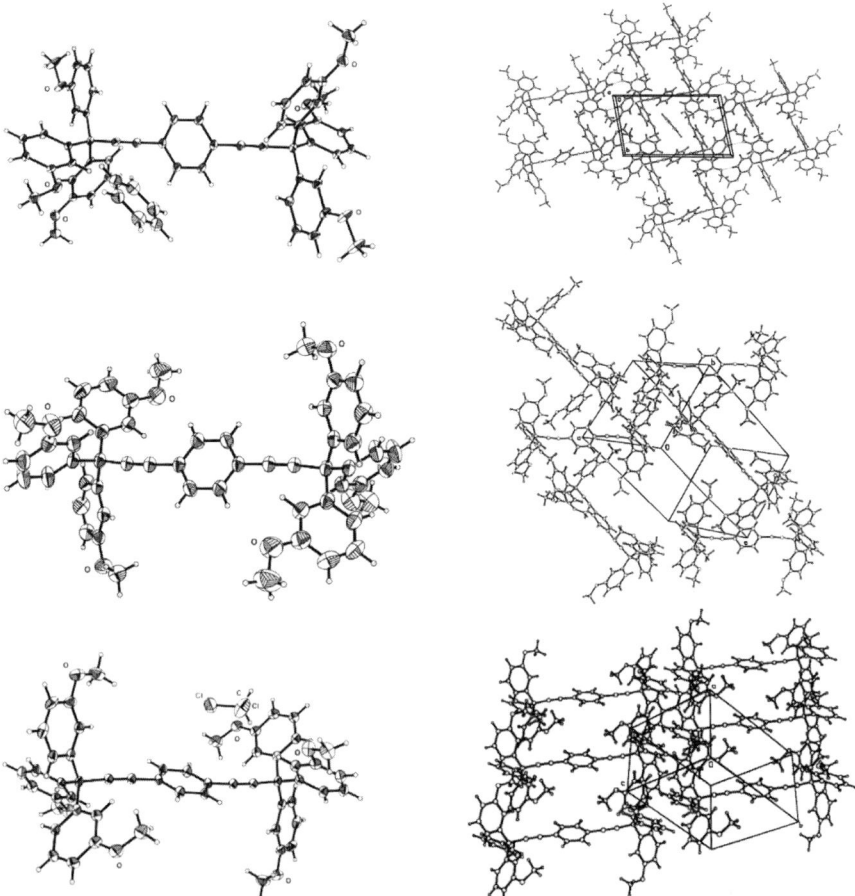

Fig. 30 (*left*) ORTEP and (*right*) packing diagrams of the hexa-*meta*-methoxy compound **6**. (*top*) Benzene clathrate (*form A*), which has only one benzene molecule at the center of the unit cell. (*middle*) Solvent-free structure (*form B*). (*bottom*) Clathrate formed from CH_2Cl_2 (*form G*) [68]

5.3
Steric Shielding Strategies in Molecular Gyroscopes with Trityl Stators

In order to facilitate the rotation of the central phenylene as compared to the parent structure, a greater degree of steric shielding will be needed. Space filling molecular models indicated that efficient steric encapsulation can be achieved with *tert*-butyl groups at the two meta positions of each of the six phenyl rings (**7**, R = *tert*-Bu) [76]. The synthesis of compound **7** was accomplished as indicated in Fig. 20 using 3,5-di-*tert*-butyl-bromobenzene as the starting material. The presence of twelve *meta-tert*-butyl groups in the structure prevents the formation of the six-fold trityl embrace seen in **2**.

X-ray analysis of single crystals from CH_2Cl_2 revealed a solvent clathrate in the monoclinic space group $P2_1/n$ with one molecule of **7** and two solvent molecules per asymmetric unit (Fig. 31). While the bulky *tert*-butyl groups in **7** provided significant steric protection, there was some interdigitation of neighboring molecules and the pocket of rotation was determined to be an elliptical cavity with cross dimensions of 5.5×8.8 Å, indicating that some distortion of the pocket would be required for the phenylene to rotate.

Even when carefully handled, crystals of **7** lost the solvent of crystallization when samples were ground for solid state NMR and thermal analysis. However, a sharp DSC melting transition at 335 °C and relatively sharp signals in the ^{13}C CPMAS spectrum suggest the desolvated phase to be crystalline, rather than amorphous (Fig. 32). The rotational dynamics of **7** were investigated by wide line ^{2}H NMR on freshly prepared crystals of samples labeled with a deuterated phenylene rotor (**7-d$_4$**, Fig. 33). Satisfyingly, spectra measured at ambient temperature (293 K) were consistent with a 2-fold flipping process in the fast exchange regime with gyroscope motion faster than 100 MHz ($k_{rot} > 10^8$ s^{-1}). It was not until samples were cooled down to 193 K that the spectrum was in the slow exchange regime with

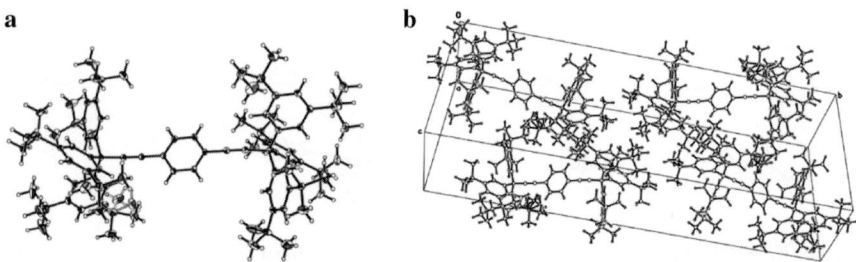

Fig. 31 **a** ORTEP diagram of molecular gyroscope **7** at 30% probability. **b** Unit cell illustrating the packing arrangement of **7** with solvent molecules excluded [76]

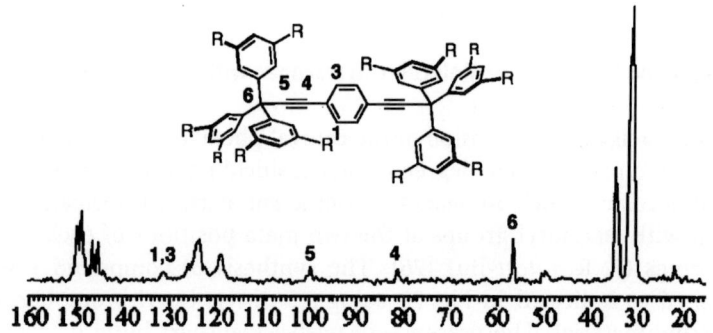

Fig. 32 ^{13}C CPMAS NMR spectrum of desolvated crystals of molecular gyroscope **7** [76]

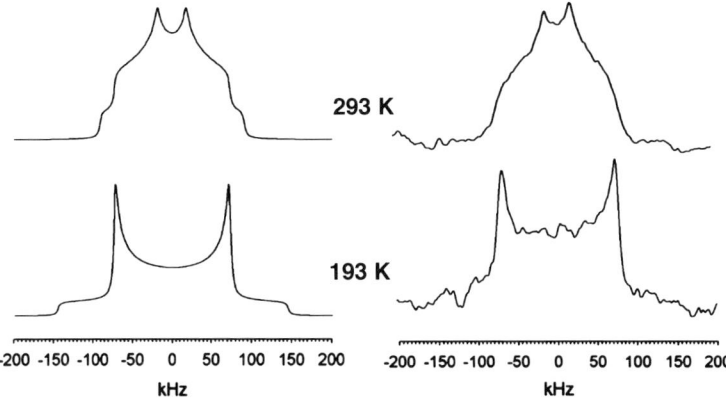

Fig. 33 Calculated (*left column*) and experimental (*right column*) quadrupolar echo solid state ^{2}H NMR of molecular gyroscope 7 [76]

$k_{rot} \leq 10^4 \text{ s}^{-1}$. Unfortunately, spectra measured at intermediate temperatures (273–213 K) revealed the fragility of the crystal lattice, as the spectra were highly heterogenous with contributions from molecules rotating in the fast exchange and the slow exchange regimes [77]. Despite the fragility of its crystal lattice, the rate of rotation in samples of freshly crystallized 7 at ambient temperature are at least five orders of magnitude greater than that of compound 2, clearly demonstrating the expected shielding effects of the *tert*-butyl groups in the substituted trityl frame and showing a path for future improvements.

6
Rotator Effects

The primary role of the stator in molecular gyroscopes is to create a shielding frame to encapsulate the rotator and protect it from environmental influences that may affect its dynamics. The stator should control also the crystallization of the system by taking advantage of its size, rigidity, aspect ratio, and functional groups at the exterior of the gyroscope frame. The role of the stator is to provide the desired dynamic functions. The dynamics of the stator should be ideally determined by its moment of inertia and rotational energy profile. Among the primary variables that can be manipulated by changing the structure of the rotator are its size, mass, and moment of inertia, its symmetry, and the presence of permanent electric or magnetic dipoles and/or quadrupoles. In Sect. 6 we will illustrate some of our early efforts at documenting the role of the rotator by analyzing a set of molecular gyroscopes with aromatic rotators, and a set of molecular compasses with substituted phenylenes.

6.1
Arylene Rotators

The suitability and generality of the Pd(0) coupling reaction for the synthesis of molecular gyroscopes with arylene rotators was investigated early on with 4,4′-biphenylene, 9,10-anthracenylene, and 3,8-pyrenylene groups. Compounds **8**, **9**, and **10** were synthesized in the same manner as compound **1** from the ethynyl triptycenes and the dihalo arylenes [52, 78].

As the space-filling models in Fig. 34 suggest, molecular rotors **1**, **8**, and **10** should be uniquely determined by the energetics of rotation about sp-sp^3 and sp-sp^2 single bonds and AM1 calculations suggest similar properties for **8** and **10** as those previously discussed for compound **1**. In contrast, because of the close proximity between the two peripheral triptycenes and the anthracene rotator in compound **9**, the energetics of rotation are expected to involve steric interactions that localize some rotational energy minima. Calculations for compound **9** indicate two structures with the triptycyl groups eclipsed, E(0), E(30), and one with the triptycyls staggered, S(30). The energies of conformers S(30) and E(30) (Figs. 35a and b) are indistinguishable from each other, and only 0.74 kcal/mol lower in energy than conformer E(0) (Fig. 35d). The highest point in the rotational surface is given by the structure of S(0) (Fig. 35c), which is only 3.93 kcal/mol above the energies of S(30) and E(30). Not surprisingly, this rotamer has the plane of the anthracene group aligned with one of the aromatic planes of each of the peripheral triptycyls, and experiences close contacts with both of them. The results with compound **1**, **8**, **9**, and **10**, are in good qualitative agreement with dynamically averaged ^{1}H NMR spectra measured in solution, which indicate a rapid conformational equilibrium, but are all expected to be static in crystals due to steric barriers from close packing interactions. Given the interesting photophysical and electrochem-

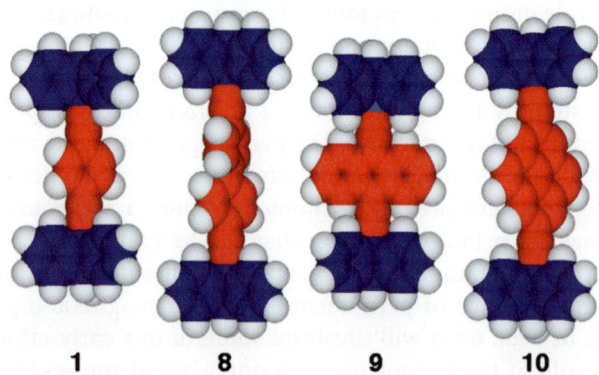

Fig. 34 Space filling models of molecular gyroscopes with phenylene-, biphenylene-, anthracenylene-, and pyrenylene-rotators

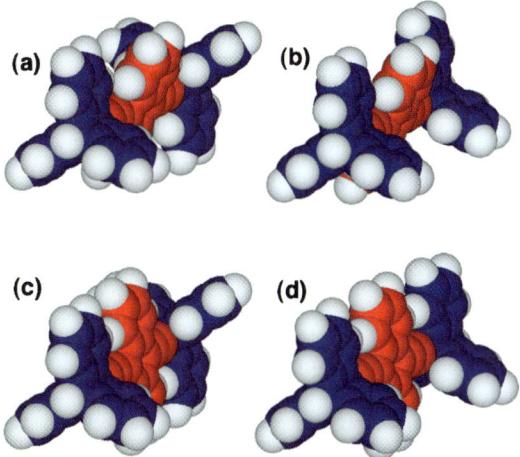

Fig. 35 Space filling models of rotamers of compound **9** [52]

ical properties of condensed aromatic systems, the design and construction of larger stators with suitable shielding groups will be highly desirable.

6.2
Phenylene Rotators with Polar Groups: Molecular Compasses

The introduction of a polar substituent orthogonal to the axis of rotation of a para-phenylene group provides opportunities to interface the internal dynamics of molecular gyroscopes with external electric, magnetic, and electromagnetic fields [79]. Given their expected response to external fields, we refer to these compounds as molecular compasses. Crystals built with molecular compasses are expected to form dipolar arrays capable of adopting spontaneous ferroelectric or antiferroelectric order depending on the symmetry of the lattice. Early work on dipolar rotators was aimed at the preparation and characterization of several polar analogs of **2** with dipole moments ranging from 0.74 to 7.30 (Fig. 36). Molecular compasses with fluoro- (**11**), cyano- (**12**), nitro- (**13**), amino- (**14**), *ortho*-diamino- (**15**) and *para*-nitroanilino- (**16**) rotators were prepared in good yields from the corresponding para-dibromo phenylenes [79].

The crystallization of non-polar aromatic molecules is generally determined by weak van der Waals forces and aromatic interactions that include π-stacking and edge-to-face interactions, which typically lead to volume-filling and close packing interactions [23, 80]. The structure of molecular gyroscope **2** can be rationalized in terms of these interactions, which result in the presence of the six-fold phenyl embrace characteristic of unsubstituted trityl groups. Barring any unusual dipole-dipole interactions that distort or completely change molecular packing, one may expect the molecular compasses **11–16** to form structures isomorphous to those of **2**, so that their

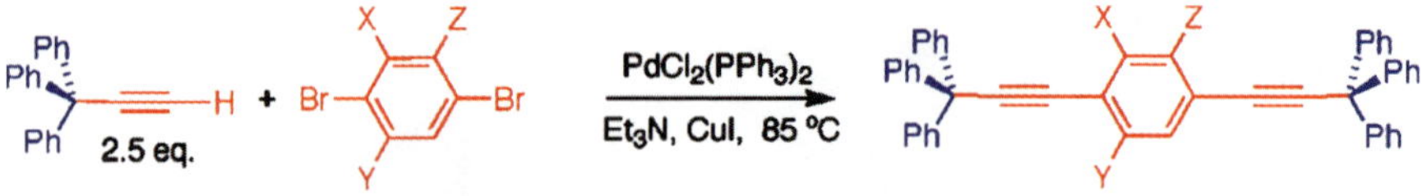

Ph–C(Ph)(Ph)–C≡C–H + Br–C₆H₂(X)(Y)(Z)–Br → [PdCl₂(PPh₃)₂, Et₃N, CuI, 85 °C] → Ph₃C–C≡C–C₆H₂(X)(Y)(Z)–C≡C–CPh₃

2.5 eq.

Molecular Compass	X	Y	Z	μ (Debye)*	Volume (Å³)
2	H	H	H	0.0	700
11	H	F	H	1.49	710
12	H	CN	H	3.01	721
13	H	NO₂	H	4.74	732
14	H	NH₂	H	1.42	706
15	NH₂	H	NH₂	0.74	720
16	H	NH₂	NO₂	7.30	736

*Dipole moments and molecular volumes were calculated from structures optimized with the AM1 method as implemented by the program Spartan.

Fig. 36 Synthesis of molecular gyroscopes with polar rotators, substitution pattern, dipole moment (μ), and molecular volume [79]

gyroscopic dynamics are expected to depend primarily on the size of the rotator as well as on potential dipole-dipole interactions.

Expectations of isomorphism were confirmed by experimental results for the parent hydrocarbon and the mono-substituted molecular compasses. With a combination of single crystal X-ray diffraction, solid state ^{13}C CPMAS NMR, and thermal analysis, it was shown that compounds **11–14** have essentially the same crystallization tendencies as compound **2** (Fig. 37). Two pseudopolymorphic crystal forms in the space group P1-bar may be obtained by crystallization from C₆H₆ and CH₂Cl₂. Crystals grown from C₆H₆ are obtained as a solvent clathrate while crystals grown from CH₂Cl₂ are solvent-free. Exceptions to the rule are the cyano derivative **12**, which formed a different crystal form in the space group $P2_1/c$ by slow evaporation from ethyl acetate, and the monoamino derivative **14**, which failed to form a benzene clathrate. Only samples of the

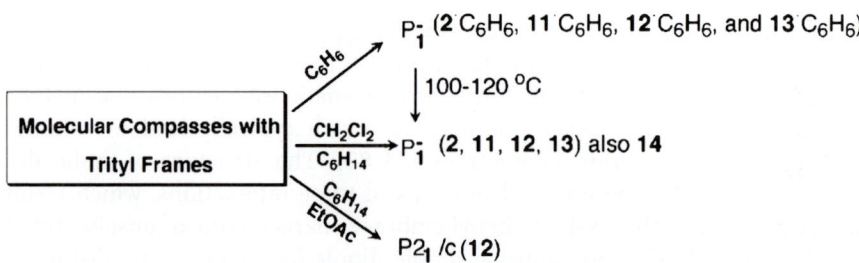

Fig. 37 Summary of crystallization properties of molecular compasses with trityl stators and polar monosubstituted phenylene rotators [79]

diamino compound **15** and the nitroanilino derivative **16** have resisted X-ray structural elucidation.

The information summarized in Fig. 37 was obtained by correlation of spectroscopic and thermal analysis with limited X-ray diffraction data. While polycrystalline samples of all compounds could be obtained from many solvents, X-ray quality single crystals were only obtained under relatively strict conditions. As expected from its isosteric similarity with its parent compound, diffraction quality single crystals of the monofluoro molecular gyroscope **11** were obtained both from benzene and from CH_2Cl_2. Crystals of the cyano derivative **12**, obtained from 3 : 1 mixtures of hexanes and ethyl acetate, were shown to be a new monoclinic form. Crystals of the amino compound **13** were obtained by slow evaporation from a 3 : 1 mixture of hexanes and CH_2Cl_2, and crystals of the nitro derivative **14** by slowly cooling hot saturated benzene solutions. Crystals of **12** grown from benzene formed a solvent clathrate analogous to those obtained from **2**, **11**, and **13**. Thermal analyses (DSC and TGA) showed all of those crystals to lose benzene at 100–120 °C, and ^{13}C CPMAS analysis carried out in situ gave spectra that were identical to those obtained from crystals grown from CH_2Cl_2.

Solvent-free crystals of fluoro (**11**) and amino (**14**) derivatives have very similar unit cells as the solvent-free structure of the parent compound (**2**)

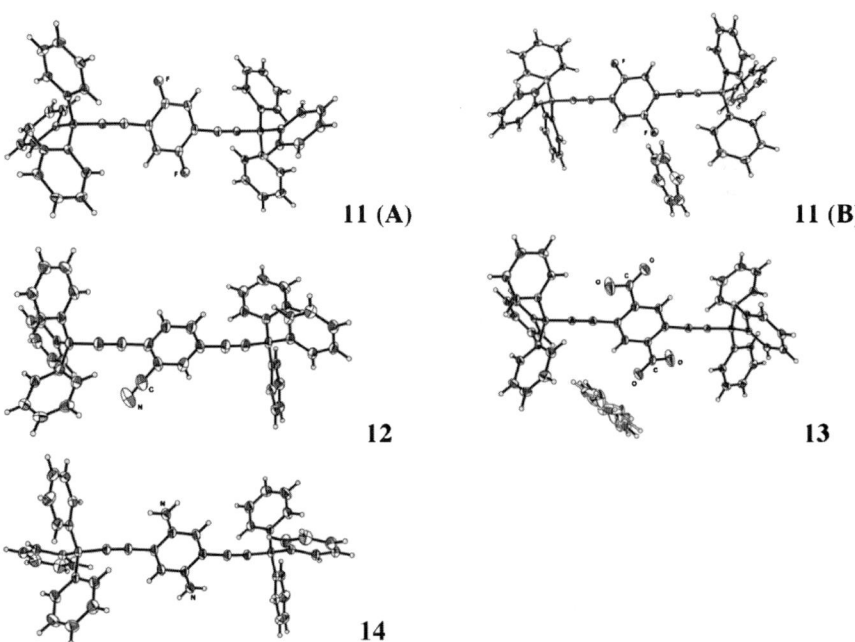

Fig. 38 ORTEP diagrams of compounds **11A**, **11B**, **12**, **13**, and **14**. Disorder in the fluoro-, amino- and nitro groups **11**, **14**, and **15**, is illustrated in the corresponding structures [79]

(Fig. 38). Similarly, the benzene clathrates formed by the fluoro (**11**) and nitro (**13**) compounds were isomorphous with benzene clathrate of **2**. Compounds **11**, **13**, and **14** showed positional disorder, with the fluoro-, nitro-, and amino-substituents having 50% occupancies in sites related by an average center of inversion in their crystal lattices. This is true for both solvent-free and benzene clathrate in the case of **11** and **14**. These observations are in contrast with the cyano derivative **12**, which crystallizes in an ordered form.

6.3
Dual Rotators: Phenylene—Diamantane

While phenylene rotators are readily accessible and ideally suited to test synthetic protocols and to set benchmarks for the dynamic behavior characteristics of a given stator, they have serious limitations as rotary units. As the shapes of rotators tend to be well-matched by their environment, flattened phenylenes tend to have cavities that result in relatively high-energy barriers and rotational potentials characterized by two energy minima connected by angular displacements of 180° and a transition state at 90° (Fig. 39). In contrast, it is expected that more cylindrical groups with higher rotational symmetry order (C_n, $n > 2$) will have energy profiles with "n" minima, 360°/n angular displacements, "n" symmetry exchangeable states, and a scaffold to build more cogs, should gearing be desirable. The effect of rotational symmetry order can also be appreciated in the figure for a set of geometries representing the cross sections of rotators (and their environments with a dotted line). The rotational symmetry order of the hypothetical rotators varies from C_1 to C_6 and C_∞. A perfectly cylindrical rotator on the right end of the figure would have an infinite number of equivalent sites and no energy barriers for rotation. In contrast, a non-symmetric C_1 rotator would only have one global minimum (and potentially, more local minima) with a rate-limiting rotational barrier that depends on how its shape departs from that of an ideal cylinder. As the symmetry order increases, as illustrated for the set of rotators with regular polygonal cross sections in Fig. 39, a better fit between the shapes of their transition states and their environments is expected. As the energy difference between minima and maxima decreases, the gyroscopic rotational rates are expected to increase.

On our way to constructing molecular gyroscopes with high symmetry rotators, we were fortunate to obtain samples of 4,9-bis(4′-iodophenyl)diamantane [81]. Appreciating that diphenyldiamantane would be an ideal test system to compare the rotational dynamics of the phenylene and the diamantanylidene groups, we set out to prepare the corresponding molecular gyroscope with a trityl stator [82]. Compound **17** was synthesized by a Pd(0)-coupling reaction of the diiododiphenyldiamantane with tritylacetylene (Fig. 40). The crystal structure of **17** was solved in space group P_{bna}

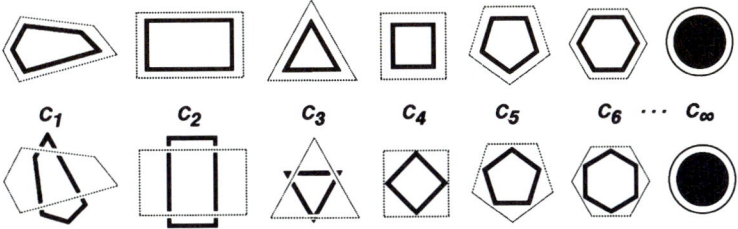

Fig. 39 Representation of the cross section of rotators with rotational axis with symmetry varying from $C_1 - C_6$ and C_∞ illustrating the topology of their environment with a *dotted line* and the increasing steric mismatch between the rotator and its environment as the symmetry order decreases

with half a molecule per asymmetric unit due to coincident molecular and crystallographic inversion centers. The packing of **17** is characterized by trityl-trityl interactions in the form of four-fold phenyl embraces, which leave the phenylene-diamantane rotators arranged in layers. It may be noted that, unlike the six-fold phenyl embrace discussed earlier, the four-fold phenyl embrace is characterized by molecules interacting in a non-colinear manner [59, 60, 82].

Very different dynamics for the two-fold phenylene and three-fold diamantane rotators were first deduced by comparison of a normal ^{13}C CPMAS spectrum and one acquired with dipolar dephasing at 25 °C (Fig. 41). The

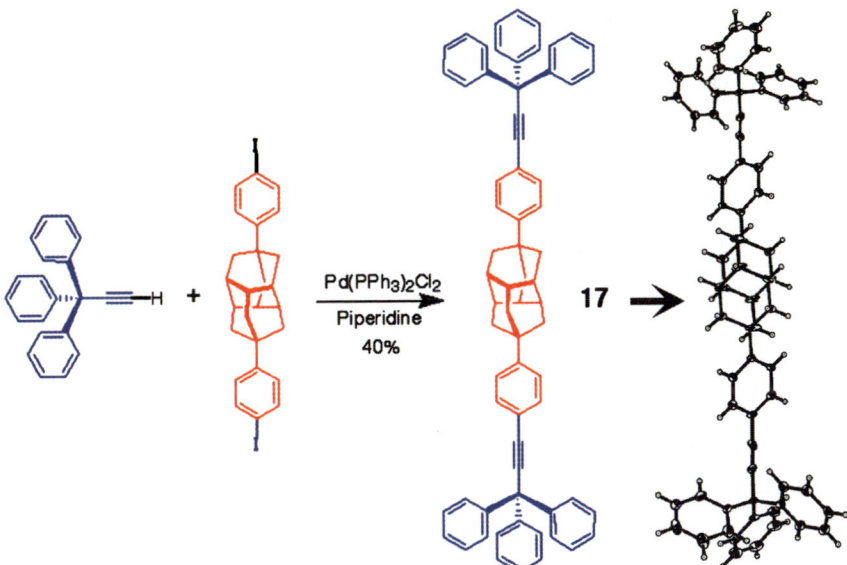

Fig. 40 Preparation and ORTEP drawing of compound **17**

nearly complete removal of the Ar-H signals near 130 ppm in the bottom spectrum is characteristic of samples with static or slow moving phenyl and phenylene groups. In contrast, the strong diamantane signals at 35–45 ppm (Diad-H) are indicative of groups that are rapidly reorienting, within the time scale of dipolar dephasing (ca. 50 μs).

To probe the rotary dynamics of phenylene we decided use variable temperature ^{13}C CPMAS NMR coalescence analysis. However, the strong overlap between the trityl and rotator signals required selective cross-polarization of the signals of interest by using short cross-polarization times (contact times) in samples of **17-d$_{30}$**, specially prepared with the trityl groups per-deuterated [82]. Variable temperature ^{13}C CPMAS of **17-d$_{30}$** carried out using contact times of only 50 μs revealed only those aromatic signals corresponding to the symmetrically related central phenylene rotators. Two sets of exchange-symmetry related signals at 123.8 and 128.2 ppm, and at 130.9 and 132.9 ppm coalesced between 240 K and 340 K, with full line shape simulations providing internally consistent exchange rate data as a function of temperature (Fig. 42). Arrhenius plots constructed with this information resulted in an activation energy of 13.7 ± 1.1 kcal/mol and a pre-exponential factor of $5.1 \pm 4.5 \times 10^{11}$ s^{-1}.

To analyze the dynamics of the central diamantylidene, ^{1}H spin-lattice relaxation measurements were conveniently carried out by indirect detection using a ^{13}C CPMAS inversion-recovery sequence. As expected for a dominant relaxation process, the results obtained indicated that all signals decay with the same mono-exponential time constant between 233 and 410 K. The relaxation data was fitted to the Kubo–Tomita expression (Fig. 43), which

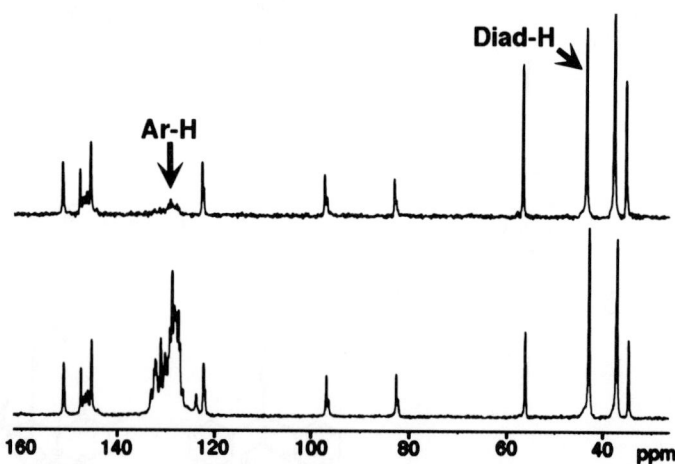

Fig. 41 (*bottom*) Normal ^{13}C CPMAS spectrum of the bisphenylenediamantane compound **17** with a contact time of 10 ms. (*top*) Spectrum of **17** with a 50 μs dipolar-dephasing [82]

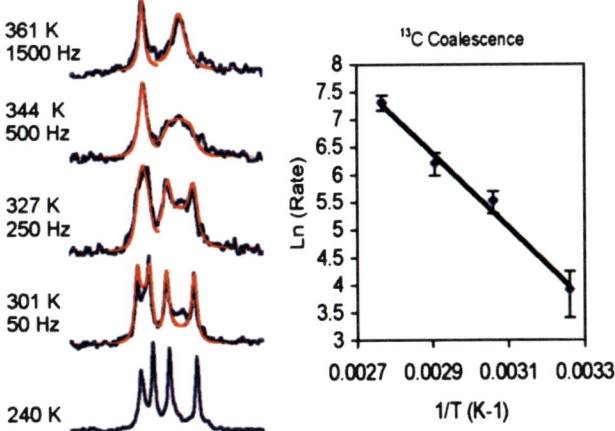

Fig. 42 (*left*) Superimposed experimental and simulated phenylene signals showing the temperature and assumed site-exchange rates. (*right*) Arrhenius plot of the coalescence data [82]

revealed an activation energy $Ea = 4.1$ kcal/mol and a pre-exponential factor $\tau_o = 2.2 \times 10^{11}$ s^{-1}. Notably, the minimum in the relaxation curve, where the inverse of the correlation time for motion, $1/\tau_c$, matches the spectrometer frequency of 300 MHz occurred at $\sim 100\,^\circ$C. Knowing that the barrier for rotation of the full diphenyldiamantylidene along the dialkyne axis is four times greater, it may be recognized that the diamantylidene must slip past the phenylene at a much faster rate. It may also be suggested that the rotational barrier for the diamantylidene group should be the sum of an intrinsic gas-phase barrier for rotation of the diamantylidene past the phenyl groups, and an environmental barrier given by the friction produced by contacts with close neighbors in the crystal. A calculation of the former with the AM1 semiempirical Hamiltonian suggested a value of 3.2 kcal/mol, which im-

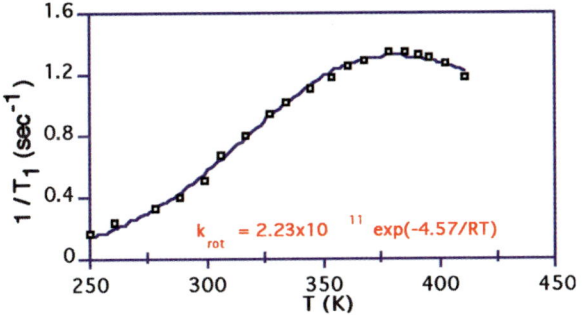

Fig. 43 Experimental results (*squares*) and curve Kubo–Tomita fit (*line*) of the spin-lattice relaxation values from compound 17 [82]

plies a solid state barrier of only ~ 1 kcal/mol. This is in stark contrast to the 10–14 kcal/mol barrier for rotation of the phenylene groups along their twofold axes, and a dramatic example of how the structure of the rotator may affect the barrier to rotation.

7
Conclusions and Outlook

By exposing a simple qualitative correlation between phase order and dynamics in condensed matter, we unveiled a poorly appreciated phase consisting of solids that combine the high positional order of a rigid crystal lattice with the fast dynamics of a liquid phase. We believe that this phase may host and support many of the mechanical and electronic functions required to develop the realm of molecular machines. Recognizing the apparently diverging requirements of this combination, we have suggested the term amphidynamic crystals or amphidynamic materials, where amphi means both sides, in analogy with other seemingly incompatible properties such as amphiphilic, amphoteric, amphibian, etc. As is the case for liquid crystals and plastic crystals, we suggested that amphidynamic crystals will require specific structural attributes to form a rigid lattice with crystalline order along with highly mobile components supported to the rigid part by molecular axles, pivots, etc.

We believe that molecular gyroscopes are a strong lead for the design and understanding of amphidynamic crystals, and we have shown several steps towards their realization. In the near future, we expect that crystalline dipolar arrays will lead to materials with promising electrooptic and dielectric functions. In fact, having confirmed that compounds with phenylene (2) and fluorophenylene (11) rotators form isomorphous crystals, we recently investigated the rotary dynamics of the monofluoro derivative 11 by ^{2}H NMR. Satisfyingly, the energy barrier for phenylene rotation in 11 (ca. 13 kcal/mol) is very similar to that of the parent compound (2). In addition, the dynamic coupling of dipolar motion with external AC fields investigated by dielectric spectroscopy reveals very similar activation parameters [83]. Many important challenges remain in the design, synthesis, crystallization, and testing of amphidynamic crystals. While much work remains in the fine-tuning of molecular compasses and gyroscopes, we have also begun work on the preparation of systems designed to respond to external photonic stimuli, and systems designed to support correlated motion. We believe that rapid progress in the last few years portends an extremely bright future for molecular machines and materials based on dynamically functional crystals.

Acknowledgements The work described in this chapter was supported by NSF grants DMR0307028, DMR0307028 (Solid state NMR), and DGE0114443 (IGERT Materials Creation Training Program). We thank Marcia Levitus (ASU), Gerardo Zepeda (UNCB IPN,

Mexico), Zaira Dominguez (U. Veracruz, Mexico), Carlos Sanrame, Hung Dang, Carlos E. Godinez, Tinh A. V. Khuong, Christopher J. Mortko, and Jose E. Nuñez for many important contributions cited here. We also thank Profs. John Price, Josef Michl, and Laura I. Clarke (now at the North Carolina State University), and Robert D. Horansky from the U. Colorado, Boulder, for on-going joint efforts, valuable discussions, and advice.

References

1. Mish FC (ed) (1991) Webster's Ninth New Collegiate Dictionary. Merriam-Webster, Inc., Springfield, MA
2. Garcia-Garibay MA (2005) Proc Nat Acad Sci 102:10771
3. McQuarrie DA, Simon JD (1997) Physical Chemistry: A Molecular Approach. University Science Books, Sausalito, CA
4. Burkert U, Allinger NL (1982) Molecular Mechanics. American Chemical Society, Washington, DC
5. Mislow K (1988) Chemtracts: Org Chem 2:151
6. Feynman RP (1960) Eng and Sci 23:22
7. Howard J (2001) Mechanics of Motor Proteins and the Cytoskeleton. Sinauer, Sunderland, MA
8. Shilwa M (ed) (2003) Molecular Motors. Wiley, Weinheim
9. Gimzewski JK, Joachim C, Schlittler RR, Langlais V, Johannsen I (1998) Science 281:531
10. Mislow K (1976) Acc Chem Res 9:26
11. Kawada Y, Iwamura H (1983) J Am Chem Soc 105:1449–1459
12. Kelly TR, Bowyer MC, Bhaskar KV, Bebbington D, Garcia A, Lang F, Kim MH, Jette MP (1994) J Am Chem Soc 116:3657
13. Bedard TC, Moore J (1995) J Am Chem Soc 107:10662
14. Balzani C, Gomez-Lopez M, Stoddart JF (1998) Acc Chem Res 31:405
15. Shima T, Hampel F, Gladysz JA (2004) Angew Chem Int Ed 43:5537
16. Kottas GS, Clarke LI, Horinek D, Michl J (2005) Chem Rev 105:1281
17. Sauvage J-P (2001) Structure and Bonding, vol 99. Springer, Berlin Heidelberg New York
18. Astumian RD (1997) Science 276:917
19. Astumian RD (2000) Phil Trans R Soc Lond B 355:511
20. Astumian RD (2001) Scientific American 285:56
21. Garcia-Garibay MA (1998) Curr Opinion in Solid State and Material Science 3/4:399
22. Dunitz JD (1979) X-Ray Analysis and the Structure of Organic Molecules. Cornell University Press, Ithaca, NY
23. Kitaigorodskii AI (1973) Molecular Crystals and Molecules. Academic Press, New York
24. Burns G (1985) Solid State Physics. Academic Press, New York
25. Zallen R (1983) The Physics of Amorphous Solids. Wiley, New York
26. Gray GW, Winsor PA (1974) Liquid Crystals and Plastic Crystals. Ellis Horwood, Chichester
27. Parsonage NG, Staveley LAK (1978) Disorder in Molecular Solids-II. Oxford University Press, Oxford, p 605–716
28. Morgan SO (1940) Ann New York Acad Sci 40:357

29. Collings PJ, Hird M (1997) Introduction to Liquid Crystals. Taylor and Francis, London
30. Orville-Thomas JW (ed) (1974) Internal Rotation in Molecules. Wiley, New York
31. Mizushima S (1954) Structure of Molecules and Internal Rotation. Academic Press, New York
32. Moro GT (1996) J Phys Chem 100:16419
33. Beer PD, Gale PA, Smith DK (1999) Supermolecular Chemistry. Oxford University Press, New York
34. Raymo FM, Stoddart JF (2001) In: Feringa BL (ed) Molecular Switches. Wiley, Weinheim, p 219–248
35. Palmer LC, Rebek JR (2004) Org Biomol Chem 2:3051
36. Hartgerink JD, Zubarev ER, Stupp SI (2002) Curr Opinion in Solid State and Material Science 5:355
37. Kottas GS, Clarke LI, Horinek D, Michl J (2005) Chem Rev 105:1281
38. Rozenbaum VM, Ogenko VM (1988) Soviet Physics-Solid State 30:1753
39. Sipachev VA, Khaikin LS, Grikina OE, Nikitin VS, Traettberg M (2000) J Mol Struct 523:1–22
40. Fyfe CA (1983) Solid State NMR for Chemists. C.F.C Press, Guelph, Ontario
41. Pines A, Gibby MG, Waugh JS (1973) J Chem Phys 59:569
42. Taylor RE (2004) Concepts Magn Reson Part A 22:37
43. Alemany LB, Grant DM, Alger TD, Pugmire RJ (1983) J Am Chem Soc 105:6697
44. Friebolin H (1998) Basic One- and Two-Dimensional NMR Spectroscopy, 3rd ed. Wiley, Weinheim
45. Line shape simulations were carried out with the program: g-NMR(2003). Adept Scientific, Bethesda
46. Hoatson GL, Vold RL (1994) NMR Basic Princ and Prog 32:1
47. Mantsch HH, Saito H, Smith ICP (1977) Progr NMR Spect 11:211
48. Greenfield MS, Ronemus AD, Vold RL, Vold RR, Ellis PD, Raidy TE (1987) J Magn Reson 72:89
49. Redfield AG (1965) Adv Mag Resonance 1:1
50. Torchia DA, Szabo A (1982) J Mag Res 49:107–121
51. Kubo R, Tomita K (1954) Phys Soc Japan 9:888
52. Godinez CE, Zepeda G, Garcia-Garibay MA (2002) J Am Chem Soc 124:4701
53. Eliel EL, Wilen SH (1994) Stereochemistry of Organic Compounds. Wiley, New York
54. Saebo S, Almlof J, Boggs JE, Stark JG (1989) Theochem 59:361
55. Abramenkov AV, Almenningen A, Cyvin BN, Cyvin SJ, Jonvik T, Khaikin LS, Roemming C, Vilkov LV (1988) Acta Chem Scand A42:674
56. Sipachev VA, Khaikin LS, Grikina OE, Nikitin VS, Traettberg M (2000) J Mol Struct 523:1
57. Godinez CE, Zepeda G, Mortko CJ, Dang H, Garcia-Garibay MA (2004) J Org Chem 69:1652
58. Godinez CE, Garcia-Garibay MA (2005) Unpublished results
59. Scudder M, Dance I (1998) J Chem Soc, Dalton Trans:329
60. Dance I, Scudder M (1998) New Journal of Chemistry 22:481
61. Dominguez Z, Dang H, Strouse MJ, Garcia-Garibay MA (2001) J Am Chem Soc 124:2398–2399
62. Sonogashira K, Tohda Y, Hagihara N (1975) Tetrahedron Lett:4467
63. Bernstein J (2002) Polymorphism in Organic Chemistry. Oxford University Press, Oxford

64. Dominguez Z, Dang H, Strouse MJ, Garcia-Garibay MA (2002) J Am Chem Soc 124:7719
65. Toda F, Ward DL, Hart H (1981) Tetrahedron Lett 22:3861
66. McNicol DD, Toda F, Bishop R (1996) Comprehensive Supramolecular Chemistry. Pergamon, Oxford
67. Karlen SD, Garcia-Garibay MA (2005) Chem Commun 189
68. Khuong T-AV, Nuñez JE, Campos LM, Farfán N, Dang H, Karlen SD, Garcia-Garibay MA (2005) Crys Growt Des (in press)
69. Corradini P (1973) Chem Ind 55:122
70. McCrone WC (1963) In: Fox D, Labes MM, Wesseberger A (eds) Physics and Chemistry of the Organic Solid State, vol 1. Wiley, New York, p 725
71. Giron D (1995) Thermochimica Acta 248:1
72. Finocchiaro P, Gust D, Mislow K (1973) J Am Chem Soc 95:8172–8173
73. Blount JF, Finocchiaro P, Gust D, Mislow K (1973) J Am Chem Soc 95:7019
74. Breen PJ, Bernstein ER, Secor HV, Seeman JI (1989) J Am Chem Soc 111:1958
75. Kumar VSS, Addlagata A, Nangia A, Robinson WT, Broder CK, Mondal R, Evans IR, Howard JAK, Alen FH (2002) Angew Chem Int Ed 41:3848–3851
76. Khuong T-AV, Zepeda G, Ruiz R, Khan SI, Garcia-Garibay MA (2004) Cryst Growth Des 4:15
77. Wehrle M, Hellmann GP, Spiess HW (1987) Colloid and Polymer Science 265:815
78. Godinez CE (2001) M.S. Thesis, University of California, Los Angeles
79. Dominguez Z, Khuong T-AV, Dang H, Sanrame CN, Nunez JE, Garcia-Garibay MA (2003) J Am Chem Soc 125:8827
80. Vainshtein BK, Fridkin VM, Indenbom VL (1982) Structure of Crystals. Springer, Berlin Heidelberg New York
81. Ortiz R (1993) Ph.D. Thesis, University of California, Los Angeles
82. Karlen SD, Ortiz R, Chapman OL, Garcia-Garibay MA (2005) J Am Chem Soc 127:6554
83. Horansky RD, Clarke LI, Price JC, Khuong T-AV, Jarowski PD, Garcia-Garibay MA (2005) Phys Rev B 72:014302

Author Index Volumes 251–262

Author Index Vols. 26–50 see Vol. 50
Author Index Vols. 51–100 see Vol. 100
Author Index Vols. 101–150 see Vol. 150
Author Index Vols. 151–200 see Vol. 200
Author Index Vols. 201–250 see Vol. 250

The volume numbers are printed in italics

Ajayaghosh A, George SJ, Schenning APHJ (2005) Hydrogen-Bonded Assemblies of Dyes and Extended π-Conjugated Systems. *258*: 83–118
Alberto R (2005) New Organometallic Technetium Complexes for Radiopharmaceutical Imaging. *252*: 1–44
Alegret S, see Pividori MI (2005) *260*: 1–36
Anderson CJ, see Li WP (2005) *252*: 179–192
Anslyn EV, see Houk RJT (2005) *255*: 199–229
Araki K, Yoshikawa I (2005) Nucleobase-Containing Gelators. *256*: 133–165
Armitage BA (2005) Cyanine Dye–DNA Interactions: Intercalation, Groove Binding and Aggregation. *253*: 55–76
Arya DP (2005) Aminoglycoside–Nucleic Acid Interactions: The Case for Neomycin. *253*: 149–178

Bailly C, see Dias N (2005) *253*: 89–108
Balaban TS, Tamiaki H, Holzwarth AR (2005) Chlorins Programmed for Self-Assembly. *258*: 1–38
Balzani V, Credi A, Ferrer B, Silvi S, Venturi M (2005) Artificial Molecular Motors and Machines: Design Principles and Prototype Systems. *262*: 1–27
Barbieri CM, see Pilch DS (2005) *253*: 179–204
Bayly SR, see Beer PD (2005) *255*: 125–162
Beer PD, Bayly SR (2005) Anion Sensing by Metal-Based Receptors. *255*: 125–162
Bier FF, see Heise C (2005) *261*: 1–25
Blum LJ, see Marquette CA (2005) *261*: 115–131
Boiteau L, see Pascal R (2005) *259*: 69–122
Boschi A, Duatti A, Uccelli L (2005) Development of Technetium-99m and Rhenium-188 Radiopharmaceuticals Containing a Terminal Metal–Nitrido Multiple Bond for Diagnosis and Therapy. *252*: 85–115
Braga D, D'Addario D, Giaffreda SL, Maini L, Polito M, Grepioni F (2005) Intra-Solid and Inter-Solid Reactions of Molecular Crystals: a Green Route to Crystal Engineering. *254*: 71–94
Brizard A, Oda R, Huc I (2005) Chirality Effects in Self-assembled Fibrillar Networks. *256*: 167–218
Bruce IJ, see del Campo A (2005) *260*: 77–111

del Campo A, Bruce IJ (2005) Substrate Patterning and Activation Strategies for DNA Chip Fabrication. *260*: 77–111

Chaires JB (2005) Structural Selectivity of Drug-Nucleic Acid Interactions Probed by Competition Dialysis. *253*: 33–53
Chiorboli C, Indelli MT, Scandola F (2005) Photoinduced Electron/Energy Transfer Across Molecular Bridges in Binuclear Metal Complexes. *257*: 63–102
Collin J-P, Heitz V, Sauvage J-P (2005) Transition-Metal-Complexed Catenanes and Rotaxanes in Motion: Towards Molecular Machines. *262*: 29–62
Collyer SD, see Davis F (2005) *255*: 97–124
Commeyras A, see Pascal R (2005) *259*: 69–122
Correia JDG, see Santos I (2005) *252*: 45–84
Costanzo G, see Saladino R (2005) *259*: 29–68
Credi A, see Balzani V (2005) *262*: 1–27
Crestini C, see Saladino R (2005) *259*: 29–68

D'Addario D, see Braga D (2005) *254*: 71–94
Davis F, Collyer SD, Higson SPJ (2005) The Construction and Operation of Anion Sensors: Current Status and Future Perspectives. *255*: 97–124
Deamer DW, Dworkin JP (2005) Chemistry and Physics of Primitive Membranes. *259*: 1–27
Deng J-Y, see Zhang X-E (2005) *261*: 171–192
Dervan PB, Poulin-Kerstien AT, Fechter EJ, Edelson BS (2005) Regulation of Gene Expression by Synthetic DNA-Binding Ligands. *253*: 1–31
Dias N, Vezin H, Lansiaux A, Bailly C (2005) Topoisomerase Inhibitors of Marine Origin and Their Potential Use as Anticancer Agents. *253*: 89–108
DiMauro E, see Saladino R (2005) *259*: 29–68
Dobrawa R, see You C-C (2005) *258*: 39–82
Du Q, Larsson O, Swerdlow H, Liang Z (2005) DNA Immobilization: Silanized Nucleic Acids and Nanoprinting. *261*: 45–61
Duatti A, see Boschi A (2005) *252*: 85–115
Dworkin JP, see Deamer DW (2005) *259*: 1–27

Edelson BS, see Dervan PB (2005) *253*: 1–31
Edwards DS, see Liu S (2005) *252*: 193–216
Escudé C, Sun J-S (2005) DNA Major Groove Binders: Triple Helix-Forming Oligonucleotides, Triple Helix-Specific DNA Ligands and Cleaving Agents. *253*: 109–148

Fages F, Vögtle F, Žinić M (2005) Systematic Design of Amide- and Urea-Type Gelators with Tailored Properties. *256*: 77–131
Fages F, see Žinić M (2005) *256*: 39–76
Fechter EJ, see Dervan PB (2005) *253*: 1–31
Fernández JM, see Moonen NNP (2005) *262*: 99–132
Fernando C, see Szathmáry E (2005) *259*: 167–211
Ferrer B, see Balzani V (2005) *262*: 1–27
De Feyter S, De Schryver F (2005) Two-Dimensional Dye Assemblies on Surfaces Studied by Scanning Tunneling Microscopy. *258*: 205–255
Flood AH, see Moonen NNP (2005) *262*: 99–132
Fujiwara S-i, Kambe N (2005) Thio-, Seleno-, and Telluro-Carboxylic Acid Esters. *251*: 87–140

Garcia-Garibay MA, see Karlen SD (2005) *262*: 179–227
Gelinck GH, see Grozema FC (2005) *257*: 135–164
George SJ, see Ajayaghosh A (2005) *258*: 83–118

Giaffreda SL, see Braga D (2005) *254*: 71–94
Grepioni F, see Braga D (2005) *254*: 71–94
Grozema FC, Siebbeles LDA, Gelinck GH, Warman JM (2005) The Opto-Electronic Properties of Isolated Phenylenevinylene Molecular Wires. *257*: 135–164
Guiseppi-Elie A, Lingerfelt L (2005) Impedimetric Detection of DNA Hybridization: Towards Near-Patient DNA Diagnostics. *260*: 161–186
Di Giusto DA, King GC (2005) Special-Purpose Modifications and Immobilized Functional Nucleic Acids for Biomolecular Interactions. *261*: 133–170

Heise C, Bier FF (2005) Immobilization of DNA on Microarrays. *261*: 1–25
Heitz V, see Collin J-P (2005) *262*: 29–62
Higson SPJ, see Davis F (2005) *255*: 97–124
Hirst AR, Smith DK (2005) Dendritic Gelators. *256*: 237–273
Holzwarth AR, see Balaban TS (2005) *258*: 1–38
Houk RJT, Tobey SL, Anslyn EV (2005) Abiotic Guanidinium Receptors for Anion Molecular Recognition and Sensing. *255*: 199–229
Huc I, see Brizard A (2005) *256*: 167–218

Ihmels H, Otto D (2005) Intercalation of Organic Dye Molecules into Double-Stranded DNA – General Principles and Recent Developments. *258*: 161–204
Indelli MT, see Chiorboli C (2005) *257*: 63–102
Ishii A, Nakayama J (2005) Carbodithioic Acid Esters. *251*: 181–225
Ishii A, Nakayama J (2005) Carboselenothioic and Carbodiselenoic Acid Derivatives and Related Compounds. *251*: 227–246
Ishi-i T, Shinkai S (2005) Dye-Based Organogels: Stimuli-Responsive Soft Materials Based on One-Dimensional Self-Assembling Aromatic Dyes. *258*: 119–160

James DK, Tour JM (2005) Molecular Wires. *257*: 33–62
Jones W, see Trask AV (2005) *254*: 41–70

Kambe N, see Fujiwara S-i (2005) *251*: 87–140
Kano N, Kawashima T (2005) Dithiocarboxylic Acid Salts of Group 1–17 Elements (Except for Carbon). *251*: 141–180
Karlen SD, Garcia-Garibay MA (2005) Amphidynamic Crystals: Structural Blueprints for Molecular Machines. *262*: 179–227
Kato S, Niyomura O (2005) Group 1–17 Element (Except Carbon) Derivatives of Thio-, Seleno- and Telluro-Carboxylic Acids. *251*: 19–85
Kato S, see Niyomura O (2005) *251*: 1–12
Kato T, Mizoshita N, Moriyama M, Kitamura T (2005) Gelation of Liquid Crystals with Self-Assembled Fibers. *256*: 219–236
Kaul M, see Pilch DS (2005) *253*: 179–204
Kaupp G (2005) Organic Solid-State Reactions with 100% Yield. *254*: 95–183
Kawasaki T, see Okahata Y (2005) *260*: 57–75
Kawashima T, see Kano N (2005) *251*: 141–180
Kay ER, Leigh DA (2005) Hydrogen Bond-Assembled Synthetic Molecular Motors and Machines. *262*: 133–177
King GC, see Di Giusto DA (2005) *261*: 133–170
Kitamura T, see Kato T (2005) *256*: 219–236
Komatsu K (2005) The Mechanochemical Solid-State Reaction of Fullerenes. *254*: 185–206

Kriegisch V, Lambert C (2005) Self-Assembled Monolayers of Chromophores on Gold Surfaces. *258*: 257–313

Lahav M, see Weissbuch I (2005) *259*: 123–165
Lambert C, see Kriegisch V (2005) *258*: 257–313
Lansiaux A, see Dias N (2005) *253*: 89–108
Larsson O, see Du Q (2005) *261*: 45–61
Leigh DA, see Kay ER (2005) *262*: 133–177
Leiserowitz L, see Weissbuch I (2005) *259*: 123–165
Lhoták P (2005) Anion Receptors Based on Calixarenes. *255*: 65–95
Li WP, Meyer LA, Anderson CJ (2005) Radiopharmaceuticals for Positron Emission Tomography Imaging of Somatostatin Receptor Positive Tumors. *252*: 179–192
Liang Z, see Du Q (2005) *261*: 45–61
Lingerfelt L, see Guiseppi-Elie A (2005) *260*: 161–186
Liu S (2005) 6-Hydrazinonicotinamide Derivatives as Bifunctional Coupling Agents for ^{99m}Tc-Labeling of Small Biomolecules. *252*: 117–153
Liu S, Robinson SP, Edwards DS (2005) Radiolabeled Integrin $\alpha_v\beta_3$ Antagonists as Radiopharmaceuticals for Tumor Radiotherapy. *252*: 193–216
Liu XY (2005) Gelation with Small Molecules: from Formation Mechanism to Nanostructure Architecture. *256*: 1–37
Luderer F, Walschus U (2005) Immobilization of Oligonucleotides for Biochemical Sensing by Self-Assembled Monolayers: Thiol-Organic Bonding on Gold and Silanization on Silica Surfaces. *260*: 37–56

Magnera TF, Michl J (2005) Altitudinal Surface-Mounted Molecular Rotors. *262*: 63–97
Maini L, see Braga D (2005) *254*: 71–94
Marquette CA, Blum LJ (2005) Beads Arraying and Beads Used in DNA Chips. *261*: 115–131
Mascini M, see Palchetti I (2005) *261*: 27–43
Matsumoto A (2005) Reactions of 1,3-Diene Compounds in the Crystalline State. *254*: 263–305
Meyer LA, see Li WP (2005) *252*: 179–192
Michl J, see Magnera TF (2005) *262*: 63–97
Milea JS, see Smith CL (2005) *261*: 63–91
Mizoshita N, see Kato T (2005) *256*: 219–236
Moonen NNP, Flood AH, Fernández JM, Stoddart JF (2005) Towards a Rational Design of Molecular Switches and Sensors from their Basic Building Blocks. *262*: 99–132
Moriyama M, see Kato T (2005) *256*: 219–236
Murai T (2005) Thio-, Seleno-, Telluro-Amides. *251*: 247–272

Nakayama J, see Ishii A (2005) *251*: 181–225
Nakayama J, see Ishii A (2005) *251*: 227–246
Nguyen GH, see Smith CL (2005) *261*: 63–91
Nicolau DV, Sawant PD (2005) Scanning Probe Microscopy Studies of Surface-Immobilised DNA/Oligonucleotide Molecules. *260*: 113–160
Niyomura O, Kato S (2005) Chalcogenocarboxylic Acids. *251*: 1–12
Niyomura O, see Kato S (2005) *251*: 19–85

Oda R, see Brizard A (2005) *256*: 167–218
Okahata Y, Kawasaki T (2005) Preparation and Electron Conductivity of DNA-Aligned Cast and LB Films from DNA-Lipid Complexes. *260*: 57–75

Otto D, see Ihmels H (2005) *258*: 161–204

Palchetti I, Mascini M (2005) Electrochemical Adsorption Technique for Immobilization of Single-Stranded Oligonucleotides onto Carbon Screen-Printed Electrodes. *261*: 27–43
Pascal R, Boiteau L, Commeyras A (2005) From the Prebiotic Synthesis of α-Amino Acids Towards a Primitive Translation Apparatus for the Synthesis of Peptides. *259*: 69–122
Paulo A, see Santos I (2005) *252*: 45–84
Pilch DS, Kaul M, Barbieri CM (2005) Ribosomal RNA Recognition by Aminoglycoside Antibiotics. *253*: 179–204
Pividori MI, Alegret S (2005) DNA Adsorption on Carbonaceous Materials. *260*: 1–36
Piwnica-Worms D, see Sharma V (2005) *252*: 155–178
Polito M, see Braga D (2005) *254*: 71–94
Poulin-Kerstien AT, see Dervan PB (2005) *253*: 1–31

Ratner MA, see Weiss EA (2005) *257*: 103–133
Robinson SP, see Liu S (2005) *252*: 193–216

Saha-Möller CR, see You C-C (2005) *258*: 39–82
Sakamoto M (2005) Photochemical Aspects of Thiocarbonyl Compounds in the Solid-State. *254*: 207–232
Saladino R, Crestini C, Costanzo G, DiMauro E (2005) On the Prebiotic Synthesis of Nucleobases, Nucleotides, Oligonucleotides, Pre-RNA and Pre-DNA Molecules. *259*: 29–68
Santos I, Paulo A, Correia JDG (2005) Rhenium and Technetium Complexes Anchored by Phosphines and Scorpionates for Radiopharmaceutical Applications. *252*: 45–84
Santos M, see Szathmáry E (2005) *259*: 167–211
Sauvage J-P, see Collin J-P (2005) *262*: 29–62
Sawant PD, see Nicolau DV (2005) *260*: 113–160
Scandola F, see Chiorboli C (2005) *257*: 63–102
Scheffer JR, Xia W (2005) Asymmetric Induction in Organic Photochemistry via the Solid-State Ionic Chiral Auxiliary Approach. *254*: 233–262
Schenning APHJ, see Ajayaghosh A (2005) *258*: 83–118
Schmidtchen FP (2005) Artificial Host Molecules for the Sensing of Anions. *255*: 1–29 Author Index Volumes 251–255
De Schryver F, see De Feyter S (2005) *258*: 205–255
Sharma V, Piwnica-Worms D (2005) Monitoring Multidrug Resistance P-Glycoprotein Drug Transport Activity with Single-Photon-Emission Computed Tomography and Positron Emission Tomography Radiopharmaceuticals. *252*: 155–178
Shinkai S, see Ishi-i T (2005) *258*: 119–160
Siebbeles LDA, see Grozema FC (2005) *257*: 135–164
Silvi S, see Balzani V (2005) *262*: 1–27
Smith CL, Milea JS, Nguyen GH (2005) Immobilization of Nucleic Acids Using Biotin-Strept(avidin) Systems. *261*: 63–91
Smith DK, see Hirst AR (2005) *256*: 237–273
Stibor I, Zlatušková P (2005) Chiral Recognition of Anions. *255*: 31–63
Stoddart JF, see Moonen NNP (2005) *262*: 99–132
Suksai C, Tuntulani T (2005) Chromogenetic Anion Sensors. *255*: 163–198
Sun J-S, see Escudé C (2005) *253*: 109–148
Swerdlow H, see Du Q (2005) *261*: 45–61
Szathmáry E, Santos M, Fernando C (2005) Evolutionary Potential and Requirements for Minimal Protocells. *259*: 167–211

Taira S, see Yokoyama K (2005) *261*: 93–114
Tamiaki H, see Balaban TS (2005) *258*: 1–38
Tobey SL, see Houk RJT (2005) *255*: 199–229
Toda F (2005) Thermal and Photochemical Reactions in the Solid-State. *254*: 1–40
Tour JM, see James DK (2005) *257*: 33–62
Trask AV, Jones W (2005) Crystal Engineering of Organic Cocrystals by the Solid-State Grinding Approach. *254*: 41–70
Tuntulani T, see Suksai C (2005) *255*: 163–198

Uccelli L, see Boschi A (2005) *252*: 85–115

Venturi M, see Balzani V (2005) *262*: 1–27
Vezin H, see Dias N (2005) *253*: 89–108
Vögtle F, see Fages F (2005) *256*: 77–131
Vögtle M, see Žinić M (2005) *256*: 39–76

Walschus U, see Luderer F (2005) *260*: 37–56
Warman JM, see Grozema FC (2005) *257*: 135–164
Wasielewski MR, see Weiss EA (2005) *257*: 103–133
Weiss EA, Wasielewski MR, Ratner MA (2005) Molecules as Wires: Molecule-Assisted Movement of Charge and Energy. *257*: 103–133
Weissbuch I, Leiserowitz L, Lahav M (2005) Stochastic "Mirror Symmetry Breaking" via Self-Assembly, Reactivity and Amplification of Chirality: Relevance to Abiotic Conditions. *259*: 123–165
Williams LD (2005) Between Objectivity and Whim: Nucleic Acid Structural Biology. *253*: 77–88
Wong KM-C, see Yam VW-W (2005) *257*: 1–32
Würthner F, see You C-C (2005) *258*: 39–82

Xia W, see Scheffer JR (2005) *254*: 233–262

Yam VW-W, Wong KM-C (2005) Luminescent Molecular Rods – Transition-Metal Alkynyl Complexes. *257*: 1–32
Yokoyama K, Taira S (2005) Self-Assembly DNA-Conjugated Polymer for DNA Immobilization on Chip. *261*: 93–114
Yoshikawa I, see Araki K (2005) *256*: 133–165
You C-C, Dobrawa R, Saha-Möller CR, Würthner F (2005) Metallosupramolecular Dye Assemblies. *258*: 39–82

Zhang X-E, Deng J-Y (2005) Detection of Mutations in Rifampin-Resistant *Mycobacterium Tuberculosis* by Short Oligonucleotide Ligation Assay on DNA Chips (SOLAC). *261*: 171–192
Žinić M, see Fages F (2005) *256*: 77–131
Žinić M, Vögtle F, Fages F (2005) Cholesterol-Based Gelators. *256*: 39–76
Zlatušková P, see Stibor I (2005) *255*: 31–63

Subject Index

Active transport 66
Amphidynamic crystals 179, 183, 185
Angular momentum 64
Anthracenes, methyl-substituted 200
Artificial molecular machines/motors 1
Arylene rotators 216
Association constants 109
ATP synthase 7

Barrier height imaging 63, 85
Barriers, rotational 88
Benzene clathrate 205
Benzylic amide rotaxanes 133
Bicyclo[2.2.2]octatriene 195
Biomolecular machines 182
Biphenyls, substituted, atropisomerism 64
Bipyridinium 12
Bistability 101
Brownian motor 93

Catenand 32
Catenanes 1, 8, 19, 29, 139
– amide-based 140
– benzylic amide 133
– rotational motion, control 159
– Ru(II) 50
Charge transfer (CT) interactions 9, 102
Clipping 104
Co-conformational changes 139
Compasses, molecular 217
Contraction/stretching 48
Copper 29, 32
CPMAS 190
Crown ether 21
Crystal dynamics/engineering 179
Cyclobis(paraquat-*p*-phenylene) cyclophane 102
Cyclophane 99

Demetallation 32
Device, nanoscale 3, 135
Diethylene glycol (DEG) 103
Dihalo arylenes 216
9,10-Dihydrophenanthrene 70
1,5-Dioxynaphthalene 19, 99, 104
Dipole flipping, field-induced 86
Donor strength, association constants 110
Driven rotation 90

Electron transfer 1
Electronic reset 18
Elevator, molecular 13
Emulating machines 185
Endergonic reactions 6
Ethane, barrier to internal rotation 64
Ethynyl triptycenes 216

Feynman, R.P. 4, 22
Field-induced dipole flipping 86

Gold(III) porphyrin, metallation-demetallation 46
Ground-state co-conformation 109
Gyroscopes, molecular 187, 215, 217
– –, triptycyl stators 195

Host-guest relationship 159

Inertial navigation systems, airplanes/satellites 187

Kinesin 7

Mercury thiolate 70
Metal sandwich stands 70
Metallation-demetallation 46
Metastable state co-conformation 109

MLCT 50
Molecular compasses 217
Molecular elevator 13
Molecular machines/motors 1, 29, 182
– –, crystalline 183
– –, electrochemical 43
– –, energy supply 5
– –, light-powered 16, 30, 50
Molecular rotors 63, 181
Molecular shuttles 32, 38, 99, 112
– –, acid-base controlled 12
Molecular switches 99, 114, 136
Molecular tinkertoys 64
Motor proteins 2
Muscle, artificial 48
Myosin 7

Nanomachines 1, 15
NMR, molecular motions 190
Nuclear rearrangements 1
Nuclear reset 18

Order, long-range 183

PAu$^+$/PZn 47
Phenanthroline 30, 40
Phenylene rotator 198
Phenylene-diamantane 220
Pirouetting, benzylic amide catenanes 143
– benzylic amide rotaxanes 144
Pirouetting machine, threaded dumbdell 41, 47
Potential energy surface 107
Proteins 2
Pseudorotaxane 17

RAM storage, catenanes 20
Reading/writing, molecular machines 6
Ring-closing metathesis (RCM) 57
RNA polymerase 7
Rotary motor, reversible 163
Rotation, driven 90
– intramolecular 64
– thermal 90

Rotational barriers 88
Rotaxanes 1, 8, 29, 38, 99, 139
– amide-based 140
– benzylic amide 133
– dimers, muscle-like 48
– ring pirouetting 157
– Ru(II) 50
Rotors, altitudinal 63, 66
– molecular 181
– surface-mounted 63, 65
Ru(diimine) 50
Ruthenium(II) 29

Self-assembly, molecular rotors 74
Shuttle, molecular 32, 38, 99, 112, 167
– –, acid-base controlled 12
Shuttling, benzylic amide rotaxanes 145
– rate 112
Spacers 113
Spin-lattice relaxation 194
Stators, triphenylmethyl 203
– triptycyl 195
Stretching/contraction 48
Superbundle 13
Supramolecular chemistry 4
Surface mounting, molecular rotors 77
Switches, molecular 99, 114, 136
– rotaxane-based 101
– two-way/three-way 159
Switching speeds 122

Tetraarylcyclobutadiene 72
4,5,9,10-Tetrahydropyrene 70
Tetrathiafulvalene 19, 99, 102, 114
Thermal rotation 90
Threading-followed-by-stoppering 104
Transition metals 30
Translational isomerism 99
Trifluoroacetate 70
Triphenylmethyl stators 203
Triptycenes 73, 195, 216
Trityl stators 203

Zinc(II) porphyrin, metallation-demetallation 46